Lecture Notes in Control and Information Sciences

For further listing of published volumes please turn over to inside of back cover.

Lecture Notes in Control and Information Sciences

Edited by A.V. Balakrishnan and M. Thoma

51

Zbigniew Nahorski
Hans F. Ravn
René V. V. Vidal

Optimization of Discrete Time Systems

The Upper Boundary Approach

Springer-Verlag Berlin Heidelberg GmbH 1983

Authors
Dr. Zbigniew Nahorski
Systems Research Inst.
The Polish Academy of Sciences
ul. Newelska 6
01-447 Warsaw
Poland

Dr. Hans F. Ravn
Danish Energy Agency
11 Landemaerket
DK-1119 Copenhagen K
Denmark

Dr. René V. V. Vidal
The Institute of Mathematical
Statistics and Operations Research
The Technical University of Denmark
DK-2800 Lyngby
Denmark

ISBN 978-3-540-12258-6   ISBN 978-3-540-39569-0 (eBook)
DOI 10.1007/978-3-540-39569-0

PREFACE

The development of numerical methods for solving optimization
problems has taken place with increasing speed along with the
development of the electronic computer. The underlying cause
for this to happen has been that a large number of technical
and operational problems have shown themselves to be suitable
for formulating and analyzing within the framework of optimization theory.

However, the speed of development has been unequal for the different classes of methods. Thus in the 1960s fundamental theoretical results were published in the area of control problems:
the Maximum Principle and the Principle of Optimality. There-
after numerical methods were developed and implemented, based on
these principles and related theorems.

In later years the main development within optimization theory
has taken place in other areas, leaving an impression that the
basic theoretical foundation for dealing with control problems
had been successfully explored and mapped. We do not share this
view.

In the present book we present a new approach to deal with the
control problems (specifically, the discrete time ones) which
indicates the possibility of exposing the fundamental theorems
not only from control theory, but also from adjoining areas of
mathematical programming. But further it provides a convenient
base for formulating new and fundamental results, in this book
exemplified by the theorem of the Nonlinear Maximum Principle.
We call the approach the Upper Boundary Approach.

While thus the book sketches some promising perspectives and
documents a number of new results, it does so, we admit, in a
preliminary form. Rather than elaborating on presentation, we
gave priority to the quick communication of the results. The
reader will hopefully agree with us in the strategy chosen!

The results presented here have their specific history in sev-
eral years of research work at IMSOR, the Institute of Mathematical Statistics and Operations Research, The Technical University
of Denmark. Here Associate Professor, Lic. Techn. René Victor
Valqui Vidal in his lectures on optimization theory and applications pointed to some problems in the perception of basic properties and concepts in the theoretical foundation of the theory.
They were basically the same in all areas of optimization theory,
and not yet quite! M.Sc. Hans Ravn, presently at the Danish
Energy Agency, at the time at IMSOR, was involved in the problems and worked with Vidal until the contours of the Upper
Boundary Approach could be seen. At this point Dr. Z. Nahorski,
of the Polish Academy of Sciences, joined the project and was
main responsible that the ideas (including some of his own) got
a firm shape at all.

We are very grateful that it was possible for the Danish Technical Research Council to support the later phase of the research economically. Also we thank M.Sc. N.O. Olesen who con-
tributed during the later phases. Last, but not least, we are
grateful to Miss Bente Wilkenschildt for her work with the preparation of this manuscript.

Copenhagen and Warsaw, July 1982

Z. Nahorski, H.F. Ravn, R.V.V. Vidal

CONTENTS

CONTENTS

CHAPTER 1

INTRODUCTION

1.0   Introduction

Optimization of discrete time systems is an activity which fre-
quently takes place as one of the central steps in the design
process, when solving certain technical problems, called multistage problems. The purpose of this activity is, generally
speaking, to find a combination of parameter values, which will
best (in some specified sense) solve the problem. To find these
values, a mathematical model is constructed, representing the
proposed solution to the problem. The search for good parameter
values is taking place as an optimization of the discrete time
model representing the multistage system.

Solution methods for the optimization of discrete time systems
is the subject of this book. We propose a new approach to this
area, the _upper boundary approach_, which will allow us to derive
new and important results, while at the same time restating
classical results within the same terminology.

This first chapter serves as an introduction to the problem area
and to the book. We shall therefore first give a number of
examples of technical and operational problems, which can be seen
to have some common properties in their mathematical representa-
tion.

A subclass of these socalled multistage optimization problems
will be identified. After a discussion and precision of this
subclass, we give a short outline of the historical development
of solution methods for it, with emphasis on Dynamic Programming
and the Maximum Principle.

1.1 Simple examples

Let us look to some problems which arise naturally in technical systems.

1° Multistage Compression of a Gas

A gas is to be isentropically compressed from the initial pressure p_0 to a final pressure p_N. The compression proceeds in N stages. In each stage the gas is first adiabatically compressed and then isobatically cooled to its initial temperature. The energy consumption at the ith stage is given by

$$E_i = mRT \ \gamma/(\gamma-1) \ [\ (x_i/x_{i-1})^{(\gamma-1)/\gamma} - 1]$$

where
m- the number of moles of gas compressed
R- the universal gas constant
T- the initial temperature of the gas
γ- the ratio of the specific heat of the gas at constant
   pressure to that at constant volume (assumed to be
   constant)
x_i-pressure of the gas at the end of the ith compression

It is desired to determine the interstage pressure for which the total energy consumed in compression is minimal. If the decision variable in the ith stage u_i is defined as

$$u_i = x_{i+1}/x_i$$

then we can formulate our problem as

$$\min_{u_0,\ldots,u_{N-1}} \sum_{i=0}^{N-1} u_i^{(\gamma-1)/\gamma}$$

subject to

$$x_{i+1} = x_i u_i \qquad\qquad i=0,1,\ldots,N-1$$

$$x_0 = p_0$$

$$x_N = p_N$$

2° Transportation Problem

Resources are to be transported from n depots (sources) to N demand points (sinks), see Fig. 1.1. It is supposed that there is only one type of resource and that the total supply is equal to the total demand.

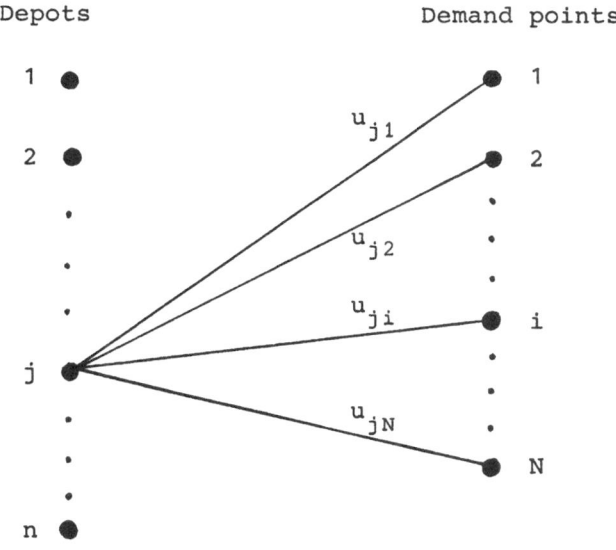

Fig. 1.1   A scheme for transportation problem.

Let

u_i^j - the quantity of the resource sent from the jth
depot to the ith demand point

$r_i^j (u_i^j)$ - the cost of this operation

The problem is to determine the quantities of u_i^j, j=1,2,...,n;
i=1,2,...,N to minimize the total costs of transporting the resources

$$R = \sum_{i=1}^{N} \sum_{j=1}^{n} r_i^j(u_i^j)$$

subject to the constraints

$$u_i^j \geq 0$$

$\sum_{i=1}^{N} u_i^j = w^j$, the supply of the resource available at the
jth depot, j=1,2,...,n

$\sum_{j=1}^{n} u_i^j = d_i$, the demand for the resource at the ith demand
point, i=1,2,...,N

j=1,2,...,n   i=1,2,...,N

We can define state variables x_i^j as the total amount of resources
which has been transported from the jth depot to the first i demand points. Then we can write the following equation

$$x_i^j = x_{i-1}^j + u_i^j$$

$$x_o^j = 0$$

$$x_N^j = w^j$$

j=1,2,...,n-1
i=1,2,...,N

It should be noted that there are only n-1 state variables although there are n depots. This arises from the fact that the demand in each demand point is preassigned. Therefore the supplies from the nth depot can be obtained by subtracting the sum of the supplies of the resource by all the rest of n-1 depots from the total demand by the ith demand point d_i i.e.

$$u_i^n = d_i - \sum_{j=1}^{n-1} u_i^j$$

It is convenient to write the above problem in vector notation. Let us then define the vectors

$$u_i = [u_i^1, u_i^2, \ldots, u_i^{n-1}]^T$$

$$x_i = [x_i^1, x_i^2, \ldots, x_i^{n-1}]^T$$

$$r_i = [r_i^1, r_i^2, \ldots, r_i^n]^T$$

$$w^j = [w^1, w^2, \ldots, w^{n-1}]^T$$

Now the problem can be formulated as

$$\min_{u_1, \ldots, u_N} \sum_{i=1}^{N} r_i(u_i)$$

subject to

$$x_i = x_{i-1} + u_i$$

$$x_o = 0$$

$$x_N = w$$

$$u_i \geq 0, \quad d_i - \sum_{i=1}^{n-1} u_i^j \geq 0$$

$$i = 1, 2, \ldots, N$$

The last two constraints define for each i a region in the n-1 dimentional control space usually referred to as a set of admissible or feasible controls.

$$\sum_{j=1}^{n} u_i^j = d_i$$

$$i=1,2,\ldots,N$$

The last five constraints define a region in the n-1 dimentional control space usually referred to as a set of admissible or feasible controls.

3° Catalyst replacement

In a catalytic reactor the efficiency of the process gradually decreases as the catalyst gets older. Because of this, the best operating conditions change in time. The problem then is to find the best operating conditions in some periods of time and the best time for replacing the catalyst so as to obtain the maximum profit. Let us consider a system depicted in Fig. 1.2 which consists of a tubular reactor and a distillation tower. In the period i a material is fed to the reactor with constant flow rate through the reactor F_i. In the reactor a compound A cracks

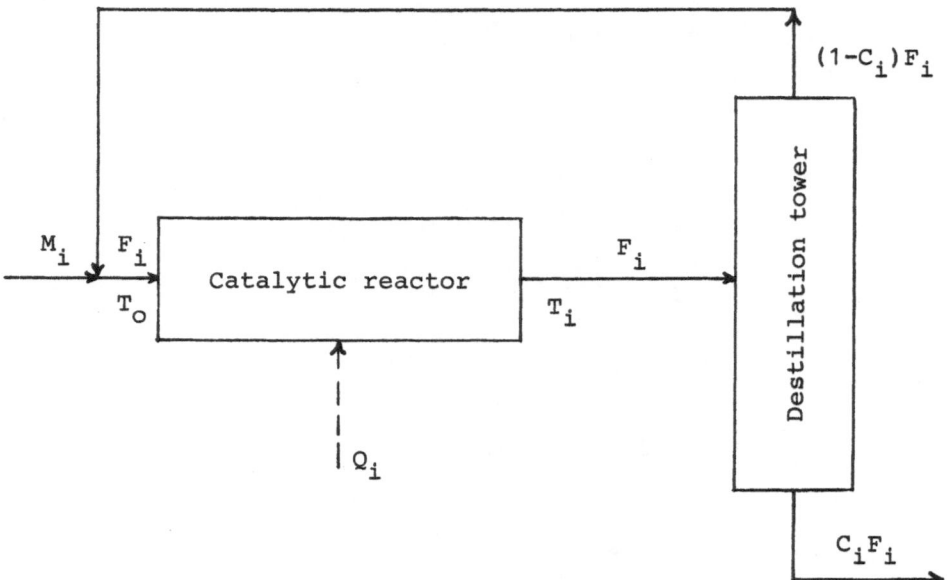

Fig. 1.2   Schematic flow sheet.

to compounds B and G. Then they go to the distillation tower from which the converted material (final product) flows with the rate C_iF_i and unconverted material is recycled with rate $(1-C_i)F_i$. The recycled material is added to the material feeded to the proces (with rate M_i) and inleted to the reactor. By material balance we have

$$M_i = C_iF_i \qquad (1.1)$$

$$F_i = M_i + (1-C_i)F_i \qquad (1.2)$$

The conversion is assumed to be expressed as

$$C_i = a_1T_i - a_2F_i - a_3S_i \qquad (1.3)$$

where T_i is the exit temperature in period i and S_i is the cumulative flow rate through the catalyst, i.e.

$$S_i = \sum_{j=1}^{i} F_i \qquad (1.4)$$

S_i represents the state of the system which is equivalent to the age of catalyst. a_1, a_2 and a_3 are given constants. By energy balance we have

$$Q_i = F_i c_p(T_i - T_o) + hC_iF_i \qquad (1.5)$$

where Q_i is the heat input to the reactor in period i, c_p is the average heat capacity of the reacting mixture, h is the heat of reaction, and T_o is the temperature of the mixture entering the reactor (assumed constant).

The conversion, temperature, and flow rates are subject to the constraints

$$C_{min} \leq C_i \leq C_{max}$$

$$T_{min} \leq T \leq T_{max}$$

$$F_{min} \leq F \leq F_{max}$$

The profit obtained per unit time is defined as

$$r_i = C_i F_i v_1 - M_i v_2 - Q_i v_3 - (1-C_i) F_i v_4 - v_5 \qquad (1.6)$$

where

v_1 - combined value of products B and G
v_2 - cost of the feed
v_3 - cost of heating
v_4 - cost of processing the recycle stream through the
distillation tower
v_5 - fixed charges.

Let us define a state variable

$$x_i = S_i$$

and the vector of decision variables

$$u_i = \begin{bmatrix} u_i^1 \\ u_i^2 \end{bmatrix} = \begin{bmatrix} T_i \\ F_i \end{bmatrix}$$

Then from equation (1.4) we have

$$x_i = x_{i-1} + u_i^2 \qquad\qquad i=1,2,\ldots,N$$

and from equations (1.1), (1.3), (1.5) and (1.6)

$$r_i(x_i, u_i) = (a_1 u_i^1 - a_2 u_i^2 - a_3 x_i) u_i^2 (v_1 - v_2 - h v_3 + v_4)$$

$$- c_p u_i^2 (u_i^1 - T_o) v_3 - u_i^2 v_4 - v_5$$

Then the problem can be formulated as follows

$$\max_{u_1,\ldots,u_N,N} \sum_{i=1}^{N} r_i(x_i,u_i)$$

subject to

$$x_i = x_{i-1} + u_i^2$$

$$x_o = 0$$

$$C_{min} \leq a_1 u_i^1 - a_2 u_i^2 - a_3 x_i \leq C_{max}$$

$$T_{min} \leq u_i^1 \leq T_{max}$$

$$F_{min} \leq u_i^2 \leq F_{max}$$

All the above examples are taken from the book by Fan & Wang (1964). The book contains references to the original papers where these problems were treated.

1.2 The multistage optimization problem

Let us try to summarize the problem formulation in the examples
from 1.1. The system is described by the equations of motions
(state equations) and the initial and/or final values of state
variables are given. The problem is to find the values of the decision variables at each stage, subject to certain constraints,
in such a way that the objective function (criterion) is maximized
or minimized. This problem is often called the multistage optimization problem.

In this paper we shall discuss only a subclass of the above defined problems. We assume that the criterion function is a sum
of the criterion functions r_i for each stage. Each of these
functions depends at most on the values of states x_i and decisions u_i in the same stage. The number of stages N is fixed and
the initial point x_o and the final point x_N of the state are
given. Only constraints on decision variables separately in
each stage are considered. Thus we assume that the problem can
be formulated as follows

$$\max \sum_{i=0}^{N-1} r_i(x_i, u_i)$$

$$x_{i+1} = g_i(x_i, u_i) \qquad\qquad (1.7)$$

$$x_o = \bar{x}_o, \; x_N = \bar{x}_N$$

$$u_i \in U_i$$

Let us discuss the limitations of this formulation in relation
to general multistage optimization problems. To do this we refer
to the 3 examples of 1.1. We see that the problem of multistage
compression of a gas is formulated as a special case of the above
problem without constraints on decision variables. The constraints form closed regions of admissible controls in the
transportation problem.

Many differences in formulation of the problem are met in the catalyst replacement problem. Let us start with constraints on the decision variables. We can observe that in this particular case the constraints are put not only on decision but also on state variables. They are of the form $h_i(u_i, x_i) \in Z_i \subset R^2 \times R^1$ – the product of the decision and state spaces. As we can always insert $x_i = g_i(x_{i-1}, u_{i-1})$, $x_{i-1} = g_{i-1}(x_{i-2}, u_{i-2})$ and so on, the state variables can always be replaced by an apropriate function of decisions in the previous stages. For the special case of catalyst replacement the function is linear and from (1.4) we have

$$x_i = \sum_{j=1}^{i} u_j^2 .$$

This way the problem with restrictions on state variables may be always reformulated as the one with restrictions on decision variables. In general they need not be, however, of the form as in problem (1.7), because they involve some constraints on functions of more than one decision vector , i.e. of $u_i, u_j, i \neq j$. Therefore the simple use of penalty functions may give better results.

The second difference in problem formulation is that the final point of the equation of motion is not fixed. Although we do not consider this problem in the paper it is relatively easy to handle it in the presented framework. The free-end-point problem, as it is usually referred to, was mainly considered in the previous works on discrete maximum principle. Sometimes, however, the problem where the final and/or initial points are not fixed but lie in some admissible regions were, studied.

The last difference in the catalyst replacement problem formulation is that the number of stages is not fixed but is subject to maximization. This kind of problems may be coped with by multiple solving of fixed-steps-number problems for different values of N.

Having discussed the limitations of the results presented in the paper and possibilities of extending the area of the applicability of the approach let us now proceed to presentation of the short history of development of the methods used to solve multistage optimization problems.

1.3 Historical notes

Many different techniques of dealing with multistage optimization problems were proposed. In special cases some heuristic methods or simply direct methods of calculation can be tried. Other cases admit application of the classical differential calculus or the calculus of variations. The problems can be also treated using linear or nonlinear programming methods. Two most specialised and yet general methods to cope with the problem were invented in late fifties. There are the Dynamic Programming and the Maximum Principle Methods.

The Dynamic Programming Method was founded and developed mainly by Bellman (1957, 1961, 1962). The method is based on the Bellman's Principle of Optimality which was formulated by Bellman (1957) as follows:

"An optimal set of decision has the property that whatever the first decision is, the remaining decisions must be optimal with respect to the outcome which results from the first decision."

Although the dynamic programming method was extended also to continuous systems it has found the main application in optimization of discrete systems or as we called them before, multistage optimization problems. The method is very powerful in treating these problems and its application is mainly limited by the extensive need for computer storage which can happen in some cases, the so called "course of dimensionality".

The Maximum Principle was first hypothesized by Pontryagin (1956) and then developed by him and his associates see Boltyanski et al (1956), Pontryagin (1957, 1959, 1961), Gamkrelidze (1957, 1958a, 1958b), Boltyanski (1958), Pontryagin et al (1962). The main idea of the maximum principle is the construction of a special function depending on controls and states and called the Hamiltonian. Knowing the optimal values of states the Hamiltonian is optimized at each stage by the optimal decision at this stage. Those first works confined to continuous systems.

The first attempt to extend the maximum principle to the optimization of discrete systems was made by Rozonoer (1959). He concluded that the maximum principle is not generally valid for discrete systems although it is valid for linear discrete systems with linear criterion. Chang (1960, 1961) discussed the validity of the maximum principle in other special types of discrete systems failing, however, to prove its applicability. This final conclusion was that the Hamiltonian may take at the optimal decisions either maximum or stationary value.

In spite of those first findings Katz (1962a, 1962b) published an incorrect derivation of the discrete maximum principle which was then quickly extended and even published in the book by Fan and Wang (1964). The first examples showing the general inapplicability of the maximum principle for discrete systems were given by Butkovski (1963, 1965) and then by Horn and Jackson (1965a, 1965b). These matters were also discussed by Denn (1965) and Gabasov (1968).

A great progress in understanding the role of the maximum principle in optimization of discrete systems was done by Halkin (1964, 1966) and Propoi (1964, 1965). They have shown that the maximum principle is valid when the set consisting of all possible values of states and criteria in each stage is convex. This assumption was then substantially weekened by Holtzman (1966a, 1966b) to directional convexity, i.e. convexity with regard to only one direction, this is the direction of increasing the criterion value.

This subject was also treated by Holtzman and Halkin (1966).

In the later works the use of nonlinear programming was stressed, see books by Canon et al. (1970) and Propoi (1973). The most popular theory at that time, the Kuhn-Tucker theory, combined with the useful approximation by cones were the basis for derivations presented in the books.

So, apart from the use of heuristic or direct methods of calculation, a number of different techniques for solving the multistage optimization problem exists. Each technique has its weak and strong sides. The dynamic programming aproach is of extremely wide applicability since it poses minimum requirements to the functions of the problem: they need neither be differential, nor continuous. In this sense it is very powerful. The price paid for this is that a lot of combinations of stages and criterion values are to be calculated and stored. Thus even fairly small problems require a great storage capacity and prove prohibitive for a solution. The maximum principle does not suffer from this weakness. To the contrary, it breaks the problem to a sequence of smaller problems. The drawback is that the functions of the problem must be well behaved in terms of continuity, differentiability and, in many practical cases, convexity or linearity.

Apparently the two methods have nothing in common unless we look to continuous-time problems. In continuous time both methods can be applied. The maximum principle even in more general cases, since the fulfilment of the requirements of directional convexity is not necessary to guarantee a solution. In continuous time the connection between the two methods is seen in e.g. the Hamilton-Jacobi equation which involve functions found both in the dynamic programming and the maximum principle methods.

On the other hand, a clear relationship between the maximum principle and nonlinear programming exists. The connecting element is the shadow price in nonlinear programming and the costate vector in the maximum principle. In numerical methods based on the penalty method the slope of the penalty function at the optimum point assumes the same value as the costate vector and the shadow price.

The reason for not using <u>nonlinear programming</u> instead of the maximum principle is obvious and well known: the number of variables involved. The strength of the maximum principle, i.e. that it reduces the number of variables at each stage of calculation, made it tempting to overcome its weaknesses. Therefore the idea of finding conditions which could be used for stage by stage optimization reducing this way the dimensionality of the problem was the main subject of other papers connected with the problem. One of them appeared in the paper by Katz (1962b) and was called the <u>weak maximum principle</u>. It stated that the optimal decision is in a stationary point of Hamiltonian. The other condition, called the <u>local maximum principle</u>, was suggested by Butkovski (1963) and then was shown to be true only for some limited class of processes. It stated that the Hamiltonian is locally maximal in an optimal decision. Then Gabasov and Kirillova (1966) proposed the <u>quasimaximum principle</u> which stated that the difference between the maximum value of Hamiltonian and the value of Hamiltonian at an optimal decision is bounded. The value of the bound can be, however, difficult to estimate.

Another approach was to generalize the Hamiltonian function in a way to keep its maximum-at-the-optimal-decision property. So called high order conditions of optimality developed by Gabasov and Tarasenko (1971) and Ashchepkov and Gabasov (1972) were the first trials in this direction. The idea of high order conditions is rather sofisticated and uses complicated formulae which makes it difficult to apply. More practically oriented generalization of the Hamiltonian was proposed by Yakovlev (1978). A new Hamiltonian was defined by adding two simple terms to the classical one.

A broader theoretical framework which leads in a natural way to generalization of Hamiltonian was elaborated by Ravn and Vidal in Ravn (1976), Vidal (1977), Ravn and Vidal (1978). Their idea was to base the derivation on the theory started by the paper by Everett (1963) and then extended by Gould (1969). The proposed theory was then sketched by Ravn (1980) in a concise form of hypothetical theorems and named the <u>Upper Boundary Method</u>.

1.4 The scope of the book

In this book we intend to show the present status of works on the
upper boundary approach to optimal control of multistage problems.
First we introduce the notion of upper boundary and discuss its
application to optimization of static (or one-stage) systems and
more specifically to the problem of maximization of a function
subject to equality constraints. This is the content of Chapter 2.
The reasoning there follows in many aspects the lines of Gould
(1969) although the definition of a support is generalized and a
saddle-point theorem for the problems with equality constraints
is introduced. Chapter 2 gives the necessary information for
understanding the derivation for multistage systems which is the
subject of Chapter 3.

In Chapter 3 the mathematical treatment of the upper boundary approach to optimization of multistage systems is presented. Many
basic notions of this chapter are the same as in Halkin (1964)
although,   to our knowledge, most of the main results connected
with the generalized maximum principle, which is presented in this
chapter, are new. In the formulation of the generalized maximum
principle one of the more restricting assumptions of the classical discrete maximum principle i.e. the assumption of directional
convexity is released. This considerably widens the area of applicability of the principle. It is shown that the classical maximum
principle is the special case of the generalized one. Apart from
the maximum principle also the dynamic programming approach is
discussed as naturally placed within the upper boundary framework.

A computer algorithm for finding controls and states satisfying
the conditions of the generalized maximum principle is proposed in
Chapter 4. This algorithm was coded and run on   a computer and the
results of computations for three simple examples give the first
evidence of the posibility to use the algorithm.

Lastly, in Chapter 5 possibilities of further research on the upper
boundary approach are sketched.

1.5 Literature

L.T. Ashchepkov, R. Gabasov: Optimization of Discrete Systems.
Differencyalnye Uravnenya, Vol. 8, No. 6, 1972 (in Russian).

R. Bellman: Dynamic Programming. Princeton Univ. Press,
New Jersey, 1957.

R. Bellman: Adaptive Control Processes. Princeton Univ. Press,
New Jersey, 1961.

R. Bellman, S.E. Dreyfus: Applied Dynamic Programming. Princeton
Univ. Press, New Jersey, 1962.

V.G. Boltyanski: The Maximum Principle in the Theory of Optimum
Processes. Doklady Akad. Nauk SSSR, Vol. 119, No. 6, 1956 (in
Russian).

V.G. Boltyanski, R.V. Gamkrelidze, L.S. Pontryagin: On the
Theory of Optimum Processes. Doklady Akad. Nauk SSSR, Vol. 110,
No. 1, 1956 (in Russian).

A.G. Butkovski: Necessary and Sufficient Conditions of Optimality
of Discrete Control Systems. Automation and Remote Control, Vol.
24, No. 8, 963-970, 1963.

A.G. Butkovski: Theory of Optimal Control of Distributed Parameter
Systems. Nauka, Moskva, 1965 (in Russian).

M.D. Canon, C.D. Cullum Jr., E. Polak: Theory of Optimal Control
and Mathematical Programming. McGraw-Hill, New York, 1970.

S.S.L. Chang: Digitized Maximum Principle. Proc. IRE, 2030-2031,
1960.

S.S.L. Chang: Synthesis of Optimum Control Systems. McGraw-Hill,
New York, 1961.

M.M. Denn: Discrete Maximum Principle. Ind. & Eng.Chem. Fundamentals, Vol. 4, No. 2, p. 240, 1965.

H. Everett III: Generalized Lagrange Multiplier Method for
Solving Problems of Optimum Allocation of Resources. Opns. Res.,
Vol. 11, No. 3, 399-417, 1963.

L.T. Fan, Ch.S. Wang: The Discrete Maximum Principle, Wiley,
New York, 1964.

R. Gabasov: Uniqueness of Optimal Control in Discrete Systems.
Izd-vo Akad. Nauk SSSR. Ser.Energetika i Avtomatika, No. 5, 1962
(in Russian).

R. Gabasov: Theory of Optimal Discrete Processes. Žurnal Vycisl.
Mat. i Mat.Fiz., Vol. 8, No. 4, 1968 (in Russian).

R. Gabasov, F.M. Kirillova: Extending L.S. Pontryagin's Maximum
Principle to Discrete Systems. Automation and Remote Control,
Vol. 27, No. 11, 1878-1882, 1966.

R. Gabasov, F.M. Kirillova: Qualitative Theory of Optimal Processes. Nauka, Moskva, 1971 (in Russian).

R. Gabasov, N.W. Tarasenko: Necessary High-Order Conditions of
Optimality for Discrete Systems. Automation and Remote Control,
Vol. 32, No. 1, 50-57, 1971.

R.V. Gamkrelidze: On the Theory of Optimum Processes in Linear
Systems. Doklady Akad. Nauk SSSR, Vol. 116, No.1, 1957 (in
Russian).

R.V. Gamkrelidze: The Theory of Time Optimal Processes in
Linear Systems. Izv. Akad. Nauk SSSR, Ser.Matem., Vol. 22,
No.4, 1958a (in Russian).

R.V. Gamkrelidze; On the General Theory of Optimum Processes.
Doklady Akad. Nauk SSSR, Vol. 123, No. 2, 1958b (in Russian).
English translation in Automation Express, Vol 1, 37-39, 1959.

F.J. Gould: Extensions of Lagrange Multipliers in Nonlinear Programming. SIAM J. Appl. Math., Vol. 17, No. 6, 1280-1297, 1969.

H. Halkin: Optimal Control for Systems Described by Difference
Equations: In: C.T. Leondes (ed): Advances in Control Systems,
Vol. I. Academic Press, 1964.

H. Halkin: A Maximum Principle of the Pontryagin Type for Systems
Described by Nonlinear Difference Equations. SIAM J. Control,
Vol. 4, No. 1, 90-111, 1966.

J.M. Holtzman: Convexity and the Maximum Principle for Discrete
Systems. IEEE Trans. Autom. Control, Vol. AC-11, No. 1, 30-35,
1966.

J.M. Holtzman: on the Maximum Principle for Nonlinear Discrete-
Time Systems. IEEE Trans. Autom. Control, Vol. AC-11, no. 2,
273-274, 1966.

J.M. Holtzman, H. Halkin: Directional Convexity and the Maximum
Principle for Discrete Systems. SIAM J. Control, Vol. 4, No. 2,
263-275, 1966.

F. Horn, R. Jackson: Discrete Maximum Principle. Ind. & Eng.
Chem. Fundamentals, Vol. 4, No. 1, 110-112, No. 4, 487-488, 1965.

R. Jackson, F. Horn: On Discrete Analogues of Pontryagin's
Maximum Principle. Int. J. Control, Vol. 1, No. 4, 389-395, 1965.

S. Katz: A Discrete Version of Pontryagin's Maximum Principle. J.
Electronics and Control, Vol. 13, p. 179, 1962.

S. Katz: Best Operating Points for Staged Systems. Ind. & Eng.
Chem. Fundamentals. Vol. 1, No. 4, 226-240, 1962.

L.S. Pontryagin: Some Mathematical Problems Arising in Connection
with the Theory of Optimum Automatic Control Systems. Session
Academy of Sciences USSR on Scientific Problems of Automating
Industry, October 15-20, 1956 (in Russian).

L.S. Pontryagin: Basic Problems of Automatic Regulation and Control. Izd-vo Akad, Nauk SSSR, 1957 (in Russian).

L.S. Pontryagin: Optimal Regulation Processes. Uspekhi Matem. Nauk, Vol. 14, No. 1, 1959 (in Russian). English translation in Am.Math. Soc. Trans., Ser. 2, Vol. 18, 321-339, 1961.

L.S. Pontryagin: Proc. First IFAC Conf., Vol. 1, p. 454, Butterworths Publishing, 1961.

L.S. Pontryagin, V.G. Boltyanski, R.V. Gamkrelidze, E.F. Mishchenko: The Mathematical Theory of Optimal Processes. Interscience, New York, 1962.

A.I. Propoi: On a Problem of Optimal Discrete Control. Doklady Akad. Nauk SSSR, Vol. 159, No. 6, 1964 (in Russian).

A.I. Propoi: The Maximum Principle for Discrete Control Systems. Automation and Remote Control, Vol. 26, No. 7, 1167-1177, 1965.

A.I. Propoi: Methods of Feasible Directions in Discrete Control Problems. Automation and Remote Controls, No. 2, 1967.

A.I. Propoi: Problems of the Discrete Control with Phase Bounds. Žurnal Vycisl. Mat. i Mat. Fiz., Vol. 12, No. 4, 1972 (in Russian).

A.I. Propoi: Elements of the Theory of Optimal Discrete Processes. Nauka, Moskva, 1973 (in Russian).

H.F. Ravn: The Discrete Maximum Principle with Nonlinear Hamiltonian. Research Notes J. nr. 2376/041928. IMSOR. Technical University of Denmark, 1976.

H.F. Ravn: Upper Boundary Methods. Research Report No. 10, IMSOR, Technical University of Denmark, 1980.

H.F. Ravn, R.V. Valqui Vidal: En elementær introduktion til
Everett's og Kuhn-Tucker's multiplikatorer og deres anvendelser.
Research Report no. 15, IMSOR, Technical University of Denmark,
1978 (in Danish).

L.I. Rozonoer: The Maximum Principle of L.S. Pontryagin in the
Theory of Optimal Systems. Automation and Remote Control, Nos
10-12, 1959.

R.V. Valqui Vidal: Notes on Static and Dynamic Optimization.
IMSOR, Technical University of Denmark, 1977.

W.M. Yakovlev: On Discrete Maximum Principle, Problemy Kibernetiki
No. 34, 1978.

CHAPTER 2

ONE STAGE SYSTEMS

0.   In this chapter we shall investigate the problem of optimization of static systems subject to equality constraints. This problem is treated using the concept of a nonlinear support to the upper boundary of the extended set of feasible solutions. The approach of this chapter is a development of the ideas presented by Everett (1963) and Gould (1969).

We start with an introduction of some basic notions and then develop the main results which consist of four theorems dealing with maximization of the socalled Hamiltonian and relations between optimal solutions to the initial problem and solutions maximizing the Hamiltonian. Neither differentiability nor continuity assumptions for the optimized function is needed. It is believed that the saddle-point theorem in this part presents a new, not earlier published result. After proving the main theorem some important cases of linear and quadratic supports are discussed more thoroughly and the chapter ends with some examples illustrating the theory and indicating some possible applications.

The chapter is written in a very short way, presenting only necessary results without any broader discussion of them. All statements are numbered so that can help in discussion of the presented results. The numbers match those in the paper by Ravn (1980). New parts which have no strict correspondence in the paper by Ravn are marked with small letters a,b,c and so on.

1. Consider the following problem:

$\textcircled{A}$

max $r(u)$ (i)

$f(u) = \bar{x}$ (ii)

$u \in U$ (iii)

where $r(\cdot)$ is a real function

$f(u)$ is a n-dimensional column vector function

u is a m-dimensional column vector (m > n)

U is a given set in R^m

$\bar{x}$ is a given vector in R^n

101. We shall work in n+1-dimensional space, $r(u)$ being measured along the first axis, and $f(u)$ along the last n axes. This space we call PR-space (P = payoff, R = resource). From properties in PR-space we shall deduce properties of the problem $\textcircled{A}$.

102. We call, for a given u,

$$x = f(u)$$

a _state_, and

$$\hat{x} = \begin{bmatrix} r(u) \\ f(u) \end{bmatrix}$$

an _extended state_.

103. If we let u to take all values in U we get a set of values of x that we shall call the _set of reachable states_, denoted by X. And likewise $\hat{x}$ will take on a set of values that we shall call the _set of reachable extended states_, denoted by $\hat{X}$.

103a.   A vector u ∈ R^m will be said to be a <u>feasible solution</u>
        for the problem Ⓐ if it satisfies (ii) and (iii) or equi-
        valently if u ∈ U and f(u) = x̄. A vector u* ∈ R^m will be
        said to be an <u>optimal solution</u> for the problem Ⓐ if it
        is feasible and r(u*) ≥ r(u), u ∈ U.

104.   For a given x we define ub(x) as the supremum of all r(u)
        for which u ∈ U and f(u) = x. ub(·) is thus defined for
        all x ∈ X.

        Considering the r(·)-axis as pointing "up" we can call
        ub(·) the supremum function or considered as a set of
        points we can call it the upper boundary of X̂, denoted
        by {ub}.

105.   We can thus conceive the problem Ⓐ as that of finding
        the u* that gives the

$$
\hat{x}* = \begin{bmatrix} r(u*) \\ \\ f(u*) \end{bmatrix} \quad \text{that on the one hand satisfies the con-}
$$

        straint   f(u) = x̄ and on the other hand lies as high up
        in X̂ as possible.

106.   <u>Theorem</u>. Ⓐ has a feasible solution u if and only if
        x̄ ∈ X and $\begin{bmatrix} ub(\bar{x}) \\ \\ \bar{x} \end{bmatrix} ∈ \hat{X}$

        <u>Proof</u>. Follows directly from the definition of a feasible
        solution and X̂.

107.   <u>Theorem</u>. If u* is an optimal solution to

        Ⓐ then
                $\hat{x}* = \begin{bmatrix} r(u*) \\ \\ f(u*) \end{bmatrix}$ is situated at {ub}

Proof. By contradiction. Suppose $\tilde{u}$ is an optimal solution and

$$\hat{\tilde{x}} = \begin{bmatrix} r(\tilde{u}) \\ \\ f(\tilde{u}) \end{bmatrix}$$

is not situated at {ub}. Then from the definition of {ub} and Theorem 106 there is a u* ∈ U for which $f(u^*) = f(\tilde{u})$ and

$$\hat{x}^* = \begin{bmatrix} r(u^*) \\ \\ f(u^*) \end{bmatrix}$$ is situated at {ub}. But then

$r(u^*) > r(\tilde{u})$ which contradicts the optimality of $\tilde{u}$.

108.   To facilitate things we shall in the sequel assume that ub(x) < ∞ for all x ∈ X and

$$\begin{bmatrix} ub(x) \\ \\ x \end{bmatrix} \in \hat{X} \text{ for all } x \in X$$

This is not a real limitation of the analysis. We shall also assume that $\bar{x} \in X$. An optimal solution we shall call u* and the corresponding extended state $\hat{x}^*$.

109.   We shall define a __support__ $\pi^o(\cdot)$ at the point x^o as a function of an n-dimensional argument, defined on the set X. It has the form

$$\pi^o(x) = \sum_{i=1}^{n} \pi_i^o(x_i)$$

It has the property that there exist a real number k, such that

(i) $\pi^o(x^o) + k = ub(x^o)$

(ii) $\pi^o(x) + k \geq ub(x)$ for all x ∈ X

We say that $\pi^o(x)$ supports $ub(x)$ at the points x for which $\pi^o(\dot{x}) + k = ub(x)$. A support at the point $\bar{x}$ we denote $\pi^*(x)$

109a. If the condition 109(ii) is replaced by (ii)' $\pi^o(x) + k > ub(x)$ for all $x \in X$ and $x \neq x^o$, we shall call a support $\pi^o(x)$ a <u>strong support</u> at a point x^o. Then it supports $ub(x)$ only at a single point x^o.

109b. If there exists a neighbourhood of x^o, $Y \subset X$ such that a function $\pi^o(x)$ satisfies conditions of §109 on the set Y then we shall call $\pi^o(x)$ a <u>local support</u> at the point x^o. If $\pi^o(x)$ satisfies on Y conditions §109 with (ii) replaced by (ii)' then we shall call $\pi^o(x)$ a <u>strong local support</u>.

109c. The definition of a support given in §109 is a generalization of a classical support definition as given by Gould (1969). The classical support is a special case of support defined in §109 subject to

$$k = 0$$

$$\pi_i^{*'}(x_i) = \lambda_i(x_i) - \lambda_i(\bar{x}_i) + \frac{1}{n} ub(\bar{x})$$

where λ_i is any function such that 109(i) and (ii) are satisfied.

Also a support obtained after a redefinition of a generalized Lagrangian as given by Evans at al (1971) is a special case of a support defined in § 109 subject to

$$k = 0$$

$$\pi_i^*(x) = \lambda_i(\bar{x}_i - x_i) + \frac{1}{n} ub(\bar{x})$$

$$\lambda_i(0) = 0$$

where λ_i, $i = 1,2,\ldots,n$ are any functions that satisfy § 109 (i), (ii), and the above conditions.

109d.   All theorems given below for supports or strong supports
        are valid for local supports or strong local supports,
        respectively, if the set X is restricted to Y.

110.   <u>Theorem</u>. If Ⓐ has a feasible solution then a $\pi^*(\cdot)$
        exists.

        <u>Proof</u>. Suppose $\bar{x}$ is a feasible value. Take

$$
\pi_i^*(x) = \begin{cases} 0 & x = \bar{x} \\ \\ \sup_{x \in X} ub(x) & x \neq \bar{x} \end{cases}
$$

        Then $\pi^*(x)$ is a support with $k = ub(\bar{x})$.

111.   <u>Theorem</u>. If $\pi^o(x)$ and $ub(x)$ are differentiable at x^o then

$$
\frac{\partial \pi^o(x^o)}{\partial x} = \frac{\partial ub(x^o)}{\partial x}
$$

        <u>Proof</u>.   Let us construct the function

$$
z(x) = \pi^o(x) - ub(x)
$$

        The function $z(x)$ is differentiable at x^o because both
        $\pi^o(x)$ and $ub(x)$ are differentiable at x^o. From the defi-
        nition of a support

$$
z(x^o) = k \quad \text{and} \quad z(x) \geq k \quad \text{for all } x \in X
$$

        then $z(x)$ has a local minimum in the point x^o
        and

$$
\frac{\partial z(x^o)}{\partial x} = \frac{\partial \pi^o(x^o)}{\partial x} - \frac{\partial ub(x^o)}{\partial x} = 0
$$

112. If ub(x) is differentiable at $\bar{x}$ we call $\dfrac{\partial ub(\bar{x})}{\partial x}$ the shadow prices.

113. Let $\pi(x)$ be any function of the form $\pi(x) = \sum\limits_{i=1}^{n} \pi_i(x_i)$. We define the Hamiltonian $H(\cdot,\cdot)$ as

$$H(u,\pi) = r(u) - \pi(f(u))$$

113a. The function $H(u,\pi)$ was previously called generalized Lagrangian, see Gould (1969). We shall call $H(u,\pi)$ the Hamiltonian in connection with dynamic programming where a generalized Hamiltonian is defined in a similar manner, see Chapter 3.

114. For given u and $\pi, H(u, \pi)$ can be interpreted as the vertical distance between the point

$$\hat{x} = \begin{bmatrix} r(u) \\ f(u) \end{bmatrix} \text{ in } \hat{X} \text{ and the point } \begin{bmatrix} \pi(f(u)) \\ f(u) \end{bmatrix} \text{ on } \pi.$$

115. **Theorem.** If u maximizes $H(u,\pi)$ over U then $\hat{x} = \begin{bmatrix} r(u) \\ f(u) \end{bmatrix}$ is situated at {ub}.

Proof. By contradiction. Suppose $\tilde{u}$ maximizes $H(u,\pi)$ over U that is

$$H(\tilde{u},\pi) \geq H(u,\pi) \quad \text{for all } u \in U$$

and $\qquad \hat{\tilde{x}} = \begin{bmatrix} r(\tilde{u}) \\ f(\tilde{u}) \end{bmatrix}$ is not situated at {ub}.

But from the definition of {ub} and assumption 108 there is a u* $\in$ U for which f(u*) = f($\tilde{u}$) and

$$\hat{x}* = \begin{bmatrix} r(u*) \\ f(u*) \end{bmatrix} \text{ is situated at } \{ub\}. \text{ Hence}$$

$r(u*) > r(\tilde{u})$ and consequently

$$H(u*,\pi) > H(\tilde{u},\pi)$$

which contradicts the supposition.

116. <u>Theorem</u>. $\pi^o(x)$ is a support at a point x^o if and only if there exists u^o which is a solution to: max $H(u,\pi^o)$ over U and satisfies $f(u^o) = x^o$.

<u>Proof</u>. Suppose $\pi^o(x)$ is a support at the point x^o.
Then from the definition of a support

$$ub(x^o) - \pi^o(x^o) = k$$
$$ub(x) - \pi^o(x) \leq k \qquad x \in X$$

Now, from the definition of ub(x) and assumption 108 there are u^o, $u \in U$ that

$$f(u^o) = x^o \text{ and } r(u^o) = ub(x^o)$$

$$f(u) = x \text{ and } r(u) = ub(x) \text{ for any } x \in X$$

Inserting the above formulae into the previous ones we have

$$r(u^o) - \pi^o(f(u^o)) = k$$
$$r(u) - \pi^o(f(u)) \leq k \quad u \in W = \{u: \begin{bmatrix} r(u) \\ f(u) \end{bmatrix} \in \{ub\}\}$$

But for any $\tilde{u} \notin W$ there is always a $u \in W$ that $r(\tilde{u}) \leq r(u)$ and $f(\tilde{u}) = f(u)$ which means that the second of above formulae is satisfied also by $u \notin W$. Thus we conclude that

$$r(u) - \pi^o(f(u)) \leq k \qquad u \in U$$

and from the definition of Hamiltonian

$$H(u^o, \pi^o) = k$$
$$H(u, \pi^o) \leq k \qquad u \in U$$

which means that u^o maximizes $H(u, \pi^o)$.

Now let u^o maximizes $H(u, \pi^o)$. Then

$$H(u^o, \pi^o) \geq H(u, \pi^o) \qquad u \in U$$

so there is a k that

$$H(u^o, \pi^o) = k$$
$$H(u, \pi^o) \leq k$$

and from the definition of Hamiltonian

$$r(u^o) - \pi^o(f(u^o)) = k$$
$$r(u) - \pi^o(f(u)) \leq k \qquad u \in U$$

But from the theorem 115 $\begin{bmatrix} r(u^o) \\ f(u^o) \end{bmatrix}$ is situated at $\{ub\}$

therefore we have $\qquad ub(x^o) - \pi*(x^o) = k$

and, defining for any $u \in U$, $f(u) = x$, we have

$$\sup_{f(u)=x} r(u) - \pi^o(f(u)) \leq k \qquad \text{all } u \in U$$

or from the definition of ub(x) and X

$$ub(x) - \pi^o(x) \leq k \qquad \text{all } x \in X$$

Rearranging terms we have

$$\pi^o(x^o) + k = ub(x^o)$$

$$\pi^o(x) + k \geq ub(x) \qquad\qquad \text{all } x \in X$$

from which we conclude that $\pi^o(x)$ is a support to $ub(x)$ at a point $x^o = f(u^o)$.

116a. <u>Corollary</u>. Let u* be an optimal solution to the problem Ⓐ. Then $\pi^*(x)$ is a support at a point $\bar{x} = f(u^*)$ if and only if u* is a solution to: max $H(u,\pi^*)$ over U.

<u>Proof</u>. As u* is a solution to the problem Ⓐ then it satisfies the condition $f(u^*) = \bar{x}$. Now application of Theorem 116 leads to 116a.

117. <u>Theorem</u>. $\pi^o(x)$ is a strong support at a point x^o if and only if any solution u^o to: max $H(u,\pi^o)$ over U satisfies $f(u^o) = x^o$.

<u>Proof</u>. The proof is a simple conclusion from Theorem 116. Suppose that $\pi^o(x)$ is a strong support at a point x^o. Then from Theorem 116 there exists u^o which maximizes $H(u,\pi^o)$ and satisfies $f(u^o) = x^o$. Suppose there exists $\tilde{u}$ which maximizes $H(u,\pi^o)$ but $f(\tilde{u}) \neq x^o$. Define $f(\tilde{u}) = \tilde{x} \neq x^o$. Then from Theorem 116 $\pi^o(x)$ is a support at the point $\tilde{x}$ which contradicts the assumption that $\pi^o(x)$ is a strong support.

Suppose now that any solution u^o to: max $H(u,\pi^o)$ over U satisfies $f(u^o) = x^o$. Then from Theorem 116 $\pi^o(x)$ is a support at x^o. Suppose it is not a strong support, i.e. that there is $\tilde{x} \neq x^o$ that $\pi^o(x)$ is a support at $\tilde{x}$. But then from Theorem 116 there exists $\tilde{u}$ which maximizes $H(u, \pi^o)$ over U and satisfies $f(\tilde{u}) = \tilde{x}$, which contradicts the assumption.

117a. <u>Corollary</u>. $\pi^*(x)$ is a strong support at a point $\bar{x}$ if and only if any solution u^* to: max $H(u,\pi^*)$ over U satisfies $f(u^*) = \bar{x}$. Moreover any such u^* is an optimal solution to the problem Ⓐ.

Proof. The first sentence of corollary is just a special case of Theorem 117. The second sentence is a conclusion from Theorem 115 and the definition of {ub}.

118. Deleted.

119. Let $\pi(x)$ be any support to $ub(x)$. Let the number k from the definition of the support be constant for all supports. Call the class of supports for given k, C_k. Define the functional $\varphi(\cdot,\cdot)$ on the product Ux C_k as

$$\varphi(u,\pi) = r(u) - \pi^*(f(u)) + \pi(\bar{x}) = H(u,\pi^*) + \pi(\bar{x})$$

120. We define a <u>saddle-point</u> of $\varphi(\cdot,\cdot)$ as a pair $(u^o,\pi^o) \in U$xC_k with the property

$$\varphi(u,\pi^o) \leq \varphi(u^o,\pi^o) \leq \varphi(u^o,\pi)$$

$u \in U$, $\pi \in C_k$, $k \in R$.

121. Let $S_k \subset C_k$ be the subset of strong supports in a set C_k. Then for $(u,\pi^o) \in U$x S_k we have:

Theorem. $\pi^*(x)$ is a strong support at the point $\bar{x} = f(u^*)$ and u^* is a solution to Ⓐ if and only if (u^*,π^*) is a saddle-point of $\varphi(\cdot,\cdot)$ restricted to Ux S_k, $k \in R$.

Proof. Let $\pi^*(x)$ be a support at the point $\bar{x} = f(u^*)$ and u^* be a solution to Ⓐ. Then from Corollary 116a

$$H(u,\pi^*) \leq H(u^*,\pi^*) \qquad \text{all } u \in U$$

Adding to both sides $\pi^*(\bar{x})$ we get the left inequality of the saddle-point definition. The right inequality we get observing that from the definition of a support

$$\pi^*(\bar{x}) \leq \pi(\bar{x}) \qquad\qquad \text{all } \pi \in S$$

Then adding to both sides $H(u^*,\pi^*)$ we get the right inequality.

Let us suppose now that (u^o,π^o) is a saddle-point of $\varphi(\cdot,\cdot)$. Then from the right inequality in the saddle-point definition we have

$$H(u^o,\pi^*) + \pi^o(\bar{x}) \leq H(u^o,\pi^*) + \pi(\bar{x}) \text{ all } \pi \in S_k$$

or

$$\pi^o(\bar{x}) \leq \pi(\bar{x}) \qquad\qquad \text{all } \pi \in S$$

Taking $\qquad \pi = \pi^*$ we have

$$\pi^o(\bar{x}) \leq \pi^*(\bar{x})$$

Suppose now that $\pi^o(x)$ is a strong support to $ub(x)$ at the point $x^o \neq \bar{x}$. Then from the definition of strong supports

$$\pi^o(\bar{x}) > \pi^*(\bar{x})$$

which contradicts the earlier inequality. This proves that $\pi^o(x)$ must be a support to $ub(x)$ at the point $\bar{x}$, i.e. $\pi^o = \pi^*$. Now, from the left inequality in the saddle-point definition

$$H(u,\pi^*) + \pi^*(\bar{x}) \leq H(u^o,\pi^*) + \pi^*(\bar{x}) \qquad \text{all } u \in U$$

or

$$H(u,\pi^*) \leq H(u^o,\pi^*) \qquad\qquad\qquad \text{all } u \in U$$

i.e. u^o maximizes $H(u,\pi*)$ over U. Then from Corollary 117a u^o is an optimal solution to the problem $\boxed{A}$, i.e. $u^o = u*$.

122.   The case of <u>linear</u> $\pi*(\cdot)$ deserves special attention . In this case

$$H(u,p) = r(u) - p^T f(u)$$

where p is a n-dimensional vector and superscript T denotes a transpose of a vector.

122a.   A nonempty set X in R^n is said to be <u>convex</u> if for any two points $x_1, x_2 \in X$ the line segment joining x_1 and x_2 is contained in X, that is, if

$$\lambda x_1 + (1-\lambda) x_2 \in X \qquad \text{for every } \lambda \in [0,1]$$

122b.   A function $f: X \to R$, with X a convex subset of R^n, is said to be <u>concave</u> if for any two points x_1 and x_2 in X and any real λ, $0 \le \lambda \le 1$ we have

$$f[\lambda x_1 + (1-\lambda) x_2] \ge \lambda f(x_1) + (1-\lambda) f(x_2)$$

123.   <u>Theorem</u>. A sufficient condition for the existence of a linear support to ub(x) at any point, $\bar{x} \in X$ is that ub(x) be finite and concave on a convex set X and $\bar{x} \in$ Int X.

<u>Proof</u>. Let us consider the set $S \subset \hat{X}$ defined in the following way

$$S = \{(x,y): x \in X, \inf_{x \in X} ub(x) \le y \le ub(x)\}$$

We prove that S is a convex set. Take any two points from S (x_1, y_1), (x_2, y_2) and construct the line segment

$$\lambda(x_1,y_1) + (1-\lambda)(x_2,y_2) = (\lambda x_1 + (1-\lambda)x_2, \quad \lambda y_1 + (1-\lambda)y_2)$$

$$0 \le \lambda \le 1$$

Now, as a set X is convex

$$x \triangleq \lambda x_1 + (1-\lambda) \, x_2 \in X$$

Then as $y_1 \le ub(x_1)$ and $y_2 \le ub(x_2)$ and $ub(x)$ is concave function, we have

$$\lambda y_1 + (1-\lambda)y_2 \le \lambda ub(x_1) + (1-\lambda) \, ub(x_2) \le ub \, [\lambda x_1 + (1-\lambda)x_2]$$

Moreover as $\inf_{x \in X} ub(x) \le y_1$ and $\inf_{y \in X} ub(x) \le y_2$

we have

$$\lambda y_1 + (1-\lambda)y_2 \ge \lambda \inf_{y \in X} ub(x) + (1-\lambda) \inf_{x \in X} ub(x) = \inf_{x \in X} ub(x)$$

Then the value

$$y \triangleq \lambda y_1 + (1-\lambda)y_2$$

satisfies

$$\inf_{x \in X} ub(x) \le y \le ub(x)$$

But this means that the point

$$(x,y) = (\lambda x_1 + (1-\lambda)x_2, \quad \lambda y_1 + (1-\lambda)y_2) \in S$$
$$\text{for any } 0 \le \lambda \le 1$$

which proves the convexity of the set S.

Then from the Theorem A.5.42 p. 251 in Canon et al (1970)
follows that the set S has supporting hyperplane L at
every boundary point $(\tilde{x}, \tilde{y}) \in S$. The hyperplane is given
by the formula

$$L = \{(x,y) \;:\; ax + a_0 y = b\}$$

where a is a n-dimensional row vector and a_0 is a scalar,
such that

$$a\tilde{x} + a_0 \tilde{y} = b$$
$$ax + a_0 y \leq b \qquad \text{for all } (x,y) \in S$$

Now, as any point $(x^0, ub(x^0))$, $x^0 \in \text{Int } X$, lies on a boundary
of S then we have

$$ax^0 + a_0 \; ub(x^0) = b$$
$$ax + a_0 \; ub(x) \leq b \quad \text{for all } x \in X$$

because $(x, ub(x)) \in S$, $x \in X$.

Let us notice, that $a_0 \neq 0$. Indeed, if $a_0 = 0$ then

$$ax^0 - b = 0 \text{ and } ax - b \leq 0 \text{ for all } x \in X$$

which means that all points $x \in X$ are on the one side of
the hyperplane $ax - b = 0$ which goes through the point x^0
Then $x^0 \notin \text{Int } X$, which contradicts the assumption.
In consequence $a_0 \neq 0$ and we have

$$cx^0 + c_0 = ub(x^0)$$
$$cx + c_0 \geq ub(x) \qquad \text{for all } x \in X$$

where

$$c = -\frac{a}{a_0} \qquad c_0 = \frac{b}{a_0}$$

From the above relation
we conclude that

$$\pi(x) = cx + c_0$$

is a linear support to ub(x) at the point x^o .

124.   In the case of a quadratic support

$$H(u,p,q) = r(u) - p^T f(u) - \tfrac{1}{2} q f^T(u) f(u)$$

where the superscript T denotes a transpose, p is a n-dimensional vector and q is a real number.

125.   <u>Theorem</u>. A sufficient condition for the existence of a quadratic support to ub(x) at any point $x^o \in X$ is that ub(x) be twice continuously differentiable in X and the function defined below of the Hessian matrix $ub_{xx}(x)$ be bounded in X, that is

$$\max_{z^T z=1} z^T ub_{xx}(x) z \le M \text{ for all } x \in X, \text{ M-real number}$$

$$z^T z=1 \qquad\qquad \text{z-real n-dimensional vector.}$$

<u>Proof</u>. Let us take q* > M and for given $x^o \in X$ take $p^{*^{\cdot}} = ub_x(x^o) - q^* x^o$. Construct the function

$$G(x,p^*,q^*) = ub(x) - p^{*T} x - \tfrac{1}{2} q^* x^T x$$

Differentiating $G(x,p^*,q^*)$ with respect to x we have

$$G_x(x^o,p^*,q^*) = ub(x^o) - p^{*T} - q^* x^o = 0$$

$$G_{xx}(x,p^*,q^*) = ub_{xx}(x) - q^* I$$

where I is the identity matrix. Let us proove that the matrix $G_{xx}(x,p^*,q^*)$ is negative definite for any $x \in X$. Indeed taking z, $z^T z \ne 0$ we have

$$y^T G_{xx} y = y^T ub_{xx}(x) y - q^* y^T y = \frac{y^T ub_{xx}(x) y}{y^T y} - q^*$$

which, putting a new vector

$$z = \frac{1}{y^T y} y \qquad\qquad z^T z = 1$$

we can write

$$y^T G_{xx} y = z^T ub_{xx}(x) z - q^* \leq \max_{z^T z = 1} z^T ub_{xx}(x) z - q^*$$

$$= \max_{z^T z = 1} z^T ub_{xx}(x) z \qquad -q^* \leq M - q^* < 0$$

which means that G_{xx} is negative definite. But now we can conclude that x^o minimizes the function $G(x, p^*, q^*)$ over X because it satisfies necessary conditions for a local maximum and moreover $G(x, p^*, q^*)$ is concave, see Theorem A. 6. 48 p. 258 in Canon at al (1970). Now from the definition of ub(x) and assumption 108 there is u^o that $f(u^o) = x^o$ and $r(u^o) = ub(x^o)$. Then u^o maximizes the Hamiltonian

$$H(u, p^*, q^*) = r(u) - p^{*T} f(u) - \tfrac{1}{2} q^* f^T(u) f(u)$$

and Theorem 116 leads to the conclusion that

$$\pi(x) = p^{*T} x + \tfrac{1}{2} q^* x^T x$$

is a support at the point x^o

126. We give two examples that quadratic function may not give a support even for quite smooth functions. Let $ub(x) = x^4$, $x \geq 0$. We try to find a quadratic support $\pi(x) = a^2 x^2 + bx + c$, $a > 0$, $k = 0$. As $ub(0) = 0$ then $c = 0$ and because both $ub(x)$ and $y(x)$ are differentiable then from Theorem III we have

$$\frac{d\ ub(0)}{dx} = 0$$

and

$$\frac{d\ \pi(0)}{dx} = b = 0$$

Now, from the definition of a support

$$ub(x) \leq \pi(x)$$

$$x^4 - a^2x^2 = x^2(x^2 - a^2) \leq 0$$
which is satisfied only for $|x| \leq a$.

Now let us take $X = <-1,1>$

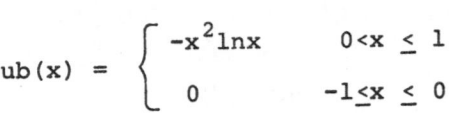 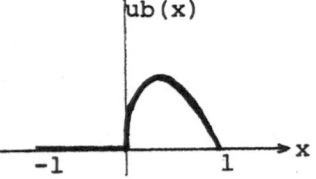

$$ub(x) = \begin{cases} -x^2 \ln x & 0 < x \leq 1 \\ 0 & -1 \leq x \leq 0 \end{cases}$$

From the de l'Hospital rule we have

$$\lim_{x \to 0^+} -x^2 \ln x = \lim_{x \to 0^+} \frac{\ln x}{-\frac{1}{x^2}} = \lim_{x \to 0^+} \frac{\frac{1}{x}}{\frac{2}{x^3}} = \lim_{x \to 0^+} \tfrac{1}{2} x^2 = 0$$

so $ub(x)$ is continuous in X. Differentiating $ub(x)$ for $x \neq 0$

$$\frac{d\ ub(x)}{dx} = \begin{cases} -2x \ln x - x & 0 < x \leq 1 \\ 0 & -1 \leq x < 0 \end{cases}$$

then from the de l'Hospital rule we have

$$\lim_{x \to 0^+} - x \ln x = 0$$

and we conclude that $\frac{d\ ub(0)}{dx} = 0$ and the fist derivative is continuous.

Let us try to find a support of the form
$\pi(x) = ax^2 + bx + c$, $k=0$. As $ub(0)=0$ and $\frac{d\ ub(0)}{dx} = 0$
then $c=0$ and $b=0$. Now from the definition of a support

$$ax^2 \geq -x^2 \ln x \qquad \text{for } 0 < x \leq 1$$
or
$$a \geq -\ln x \qquad \text{for } 0 < x \leq 1$$

which is impossible because $\lim\limits_{x \to 0^+} (-\ln x) = \infty$.

127. Below we give examples of finding the direction for changing supports.

Example 1. Linear support. Suppose that we assumed a support $\pi(x) = a^T x + b$ and we found that $\tilde{u}$, $f(\tilde{u}) = \tilde{x}$, maximizes the Hamiltonian. If $\pi^*(x) = a^{*T}x + b^*$ is a support at the point $\bar{x}$ then from the definition of supports (taking $k=0$)

$$\pi(\tilde{x}) \leq \pi^*(\tilde{x}) \qquad\qquad \pi(\bar{x}) \geq \pi^*(\bar{x})$$

which leads to the condition

$$(a^* - a)^T(\bar{x} - \tilde{x}) \leq 0$$

For the one-dimensional problem we have

$$a^* < a \qquad \text{for } \bar{x} > \tilde{x}$$
$$a^* > a \qquad \text{for } \bar{x} < \tilde{x}$$

which is illustrated on the figures below

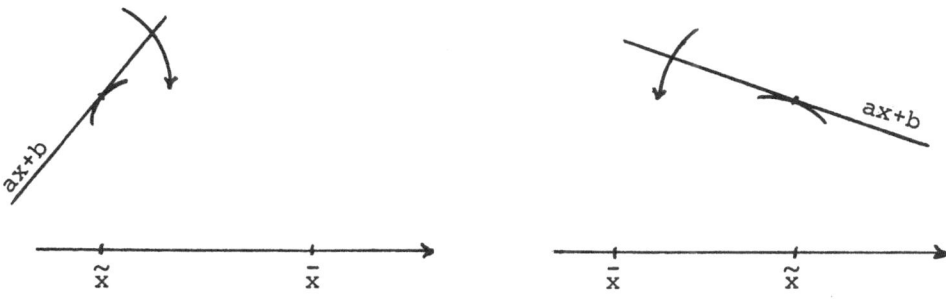

In n-dimensional case $n>1$ any choice of a^* satisfying the condition $(a^*-a)^T(\bar{x}-\tilde{x}) \leq 0$ should give better results (if $||a^*-a||$ is not too big) but the optimal direction is given by $a^* - a = -(\bar{x}-\tilde{x})$ which leads to the algorithm
$$a_{k+1} = a_k + \beta(\bar{x}-\tilde{x}), \quad \beta < 0 .$$

<u>Example 2</u>. Quadratic support. Suppose that for a support $\pi(x) = a\, x^T x + b^T x + c$ we found a point $\tilde{u}$, $f(\tilde{u}) = \tilde{x}$, which maximizes the Hamiltonian. Then, if $\pi^*(x) = a^* x^T x + b^{*T} x + c^*$ is a support at the point $\bar{x}$, we have as before

$$\pi(\tilde{x}) \leq \pi^*(\tilde{x}) \qquad\qquad \pi(\bar{x}) \geq \pi^*(\bar{x})$$

which leads to the condition

$$(a^*-a)(\bar{x}^T\bar{x} - \tilde{x}^T\tilde{x}) + (b^*-b)^T(\bar{x}-\tilde{x}) \leq 0$$

If we fix $a^*=a$ then the condition takes the form

$$(b^*-b)^T(\bar{x}-\tilde{x}) \leq 0$$

For the one-dimensional case we have

$$b^* < b \qquad \text{for } \bar{x} > \tilde{x}$$
$$b^* > b \qquad \text{for } \bar{x} < \tilde{x}$$

which can be illustrated as on the figures below.

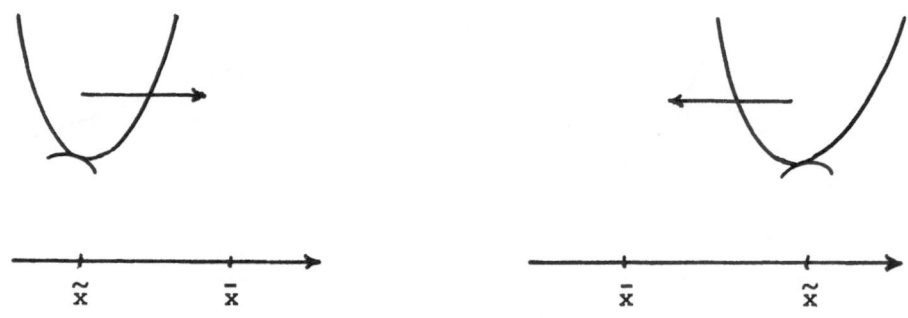

After the same argumentation as in Example 1 we conclude that the adjustment of parameters b should take form

$$b_{k+1} = b_k + \beta(\tilde{x} - \bar{x}), \quad \beta < 0 .$$

0. In this chapter we consider a problem of dynamic optimization
of discrete systems. The problem presented is the so called
fixed-endpoint problem. To solve this kind of problems mainly
dynamic programming algorithms were tried up to now. These
algorithms are very succesful unless they meet the problem
of too many dimensions.   The other methods based
on the maximum principle were rather applied to the free-end-
point problems. The new approach given in this chapter provides
a broader theoretical fremework within which both dynamic programming and maximum principle can be included. More specifically the argumentation of this chapter leads to the generalized maximum principle. It requires much less restricting
assumption to apply it than the classical maximum principle
and the classical one is a special (linear) version of it.
Then the forward and backward version of the dynamic programming are fitted in the framework. Many results of this chapter
are new ones especially those concerning the generalized maximum principle.

The argumentation of this chapter relies heavily on the results
developed in chapter 2. The reason is that when using the maximum principle the overall optimization problem is split to a
number of subproblems where the ideas of chapter 2 may be readi-
ly applied. The reformulated chapter 2 theorems constitute a
significant part of the present chapter.

Like in chapter 2 the statements of this chapter are numbered.
The numbers match those of the paper by Ravn (1980).

2. Consider the following problem Ⓑ :

$$\max \sum_{i=0}^{N-1} r_i(x_i, u_i)$$

$$x_{i+1} - x_i = f_i(x_i, u_i)$$

$$u_i \in U_i$$

$$x_0 = \bar{x}_0 , \quad x_N = \bar{x}_N$$

where $r_i(\cdot, \cdot)$ is a real function

$f_i(\cdot, \cdot)$ is a n-dimensional column vector function

u_i is a m-dimensional column vector

x_i is a n-dimensional column vector

U_i is a given set in R^m

$\bar{x}_0$ and $\bar{x}_N$ are given vectors

for $i = 0, \cdots, N-1$.

201. We shall work in $n+1$ dimensional space, where the x_i are measured along the last n axes, and where the accumulated criterion

$$R_i(u_0, u_1, \cdots, u_{i-1}) = \sum_{j=0}^{i-1} r_j(x_j, u_j) \quad i = 1, \cdots, N$$

is measured along the first axis. This space we call PR-space.

202. We call the x_i <u>states</u>, and we define

$$\hat{x}_i = \begin{bmatrix} R_i(\cdot, \cdot, \cdots, \cdot) \\ x_i \end{bmatrix}$$

as <u>extended states</u>.

203. If we let u_0 take all values in U_0 , we shall see that x_1 take on a set of values, that we shall call the <u>set of</u>

<u>states reachable from</u> $\bar{x}_0$, denoted $X_1(\bar{x}_0)$ or X_1 .

Likewise $\hat{x}_1$ will take on a set of values, that we shall call the <u>set of extended states reachable from</u> $\hat{\bar{x}}_0 = \begin{bmatrix} 0 \\ \bar{x}_0 \end{bmatrix}$, denoted $\hat{X}_1(\hat{\bar{x}}_0)$ or $\hat{X}_1$.

204. In accordance with the analysis of Ⓐ we shall assume that the upper boundary is included in $\hat{X}_1(\hat{\bar{x}}_0)$. We shall denote this upper boundary by

$$ub_1(\hat{\bar{x}}_0 , x_1) \qquad \text{and} \qquad UB_1(x_1)$$

conceived as functions of x_1 , and

$$\{ub(\hat{\bar{x}}_0)\} \qquad \text{and} \qquad \{UB_1\}$$

conceived as sets.

205. From a given $\hat{x}_1 = \begin{bmatrix} R_1 \\ x_1 \end{bmatrix}$ in X_1 we let u_1 assume all values in U_1 , and we thus generate the <u>set of states reachable from</u> x_1 $X_2(x_1)$, the <u>set of extended states reachable from</u> $\hat{x}_1$ $\hat{X}_2(\hat{x}_1)$, and the <u>(smaller) upper boundary</u> $ub_2(\hat{x}_1 , \hat{x}_2)$ or $\{ub_2(\hat{x}_1)\}$.

206. This we do for all $\hat{x}_1$ in $\hat{X}_1$. We thus generate the <u>set of reachable states</u> X_2 (i.e. reachable from $\bar{x}_0$), the <u>set of extended reachable states</u> $\hat{X}_2$ and the <u>(greater) upper boundary</u> $UB_2(x_2)$ or $\{UB_2\}$.

207. This can be generalized to any stage i . Note the following characteristics:

(i) $ub_1(\hat{\bar{x}}_0 , x_1) = UB_1(x_1)$ or $\{ub_1(\hat{\bar{x}}_0)\} = \{UB_1\}$

(ii) $ub_i(\hat{x}_{i-1} , x_i) \leq UB_i(x_i)$ $\quad \forall x_i \in X_i(x_{i-1})$

208. We shall assume that the upper boundaries are all included in the proper sets of reachable extended states. And that $x_N \in X_N$. An optimal solution we call u_i^*, i = 0,1,$\cdots$, N-1 , with the corresponding states and extended states x_i^* and $\hat{x}_i^*$, i = 1,2,$\cdots$,N , respectively.

208a. Let us notice that from the definitions of extended states in § 202 and criterion in § 201 the extended states satisfy the difference equation

$$\hat{x}_{i+1} = \hat{x}_i + F_i(\hat{x}_i, u_i)$$

where

$$F_i(\hat{x}_i, u_i) = \begin{bmatrix} r_i(x_i, u_i) \\ f_i(x_i, u_i) \end{bmatrix}$$

209. <u>Theorem.</u> $\hat{x}_i^*$ is situated at $\{UB_i\}$, i = 1,2,$\cdots$,N .

This is Halkin's Principle of Optimal Evolution, Halkin (1964).

<u>Proof.</u> We start from the stage N. Let us assume that

$$\hat{x}_N^* = \begin{bmatrix} R_N^* \\ x_N^* \end{bmatrix}$$

is not situated at $\{UB_N\}$. Then from the definition of $\hat{X}_N$ and assumption 208 there exist a control sequence $\{\tilde{u}_0, \tilde{u}_1, \cdots, \tilde{u}_{N-1}\}$ that the corresponding point

$$\hat{\tilde{x}}_N = \begin{bmatrix} \tilde{R}_N \\ \tilde{x}_N \end{bmatrix}$$

satisfies

$$\tilde{R}_N > R_N^* \qquad \tilde{x}_N = x_N^*$$

which contradicts the optimality of $\hat{x}_N^*$.

Let $\{u_0^*, u_1^*, \cdots, u_{N-1}^*\}$ and $\{x_1^*, x_2^*, \cdots, x_N^*\}$ be optimal sequences of controls and states. Assume that for $1 \leq i \leq N-1$

$$\hat{x}_i^* = \begin{bmatrix} R_i^* \\ \\ x_i^* \end{bmatrix}$$

is not situated at $\{UB_i\}$. Then from the definition of $\hat{x}_i$ and assumption 208 there exist a control sequence $\{\tilde{u}_0, \tilde{u}_1, \cdots, \tilde{u}_{i-1}\}$ that the corresponding point

$$\hat{\tilde{x}}_i = \begin{bmatrix} \tilde{R}_i \\ \\ \tilde{x}_i \end{bmatrix}$$

satisfies

$$\tilde{R}_i > R_i^* \qquad \tilde{x}_i = x_i^*$$

Apply now the control sequence

$$\{\tilde{u}_0, \tilde{u}_1, \cdots, \tilde{u}_i, u_{i+1}^*, \cdots, u_N^*\}$$

Above control sequence is admissible, that is it satisfies the constraints $u_i \in U_i$. It is now easily seen that the above control sequence produces the value of the objective function

$$\tilde{R}_N = \tilde{R}_i + \sum_{j=i}^{N-1} r_j\left(x_j^*, u_j^*\right)$$

But the value of the optimal objective function is equal to

$$R_N^* = R_i^* + \sum_{j=i}^{N-1} r_j\left(x_j^*, u_j^*\right)$$

and we have

$$\tilde{R}_N > R_N^*$$

which contradicts the optimality of $\{u_0^*, u_1^*, \cdots, u_{N-1}^*\}$

210. **Theorem.** $\hat{x}_i^*$ is situated at $\{ub_i(x_{i-1}^*)\}$, $i = 1, 2, \cdots, N$.

Proof. Let $\{x_1^*, x_2^*, \cdots, x_N^*\}$ be a sequence of optimal states. We have

$$\hat{x}_i^* = \begin{bmatrix} R_i^* \\ \\ x_i^* \end{bmatrix}$$

From the definition of (smaller) upper boundary

$$R_i^* \leq ub_i(\hat{x}_{i-1}^*, x_i^*)$$

and from the Theorem 209

$$R_i^* = UB_i(x_i^*)$$

Now from 207 (ii)

$$ub_i(\hat{x}_{i-1}^*, x_i^*) \leq UB_i(x_i^*) = R_i^*$$

Then we conclude

$$R_i^* = ub_i(\hat{x}_{i-1}^*, x_i^*)$$

that is $\hat{x}_i^*$ is situated at $\{ub_i(\hat{x}_{i-1}^*)\}$.

211. Deleted.

212. We define a support $\pi_{i+1}^o(x_{i+1})$ to $\{ub_{i+1}(\hat{x}_i)\}$ at a point x_{i+1}^o as a function of a n-dimensional argument, defined on the set $X_{i+1}(x_i)$. It has the form

$$\pi_{i+1}^o(x_{i+1}) = \sum_{j=1}^{n} \pi_{i+1}^j(x_{i+1}^j)$$

It has the property that there exists a real number k_{i+1} such that

$$\pi^*_{i+1}\left(x^*_{i+1}\right) + k_{i+1} = ub_{i+1}\left(\widehat{x^*_i}, x^*_{i+1}\right)$$

$$\pi^*_{i+1}\left(x_{i+1}\right) + k_{i+1} \geq ub_{i+1}\left(\widehat{x^*_i}, x_{i+1}\right) \qquad \forall x_{i+1} \in X_{i+1}\left(x^*_i\right)$$

Finally we require that $\pi^*_{i+1}\left(x^*_{i+1}\right)$ be finite.

213. <u>Theorem</u>. If Ⓑ has a solution, then at least one $\pi^*_i(\cdot)$ exists, $i = 1, 2, \cdots, N$.

<u>Proof</u>. Suppose u^*_i , $i = 0, 1, \cdots, N-1$, is a solution to Ⓑ , with x^*_i , $i = 1, 2, \cdots, N$ being the corresponding states. For any i take

$$\pi^*_i\left(x_i\right) = \begin{cases} 0 & x_i = x^*_i \\[2ex] \sup\limits_{x_i \in X_i\left(x^*_{i-1}\right)} ub_i\left(x^*_{i-1}, x_i\right) & x_i \neq x^*_i \end{cases}$$

Then $\pi^*_i\left(x_i\right)$ is a support to $ub_i\left(x^*_{i-1}, x_i\right)$ at the point x^*_i with $k = ub_i\left(x^*_{i-1}, x^*_i\right)$.

214. We define the Hamiltonian $H_i\left(x_i, u_i, \pi_{i+1}\right)$ at stage i as

$$H_i\left(x_i, u_i, \pi_{i+1}\right) = r_i\left(x_i, u_i\right) - \pi_{i+1}\left[x_i + f_i\left(x_i, u_i\right)\right]$$

214a. The above definition is a generalization of definitions of Hamiltonian for the case of the linear support. However, in the earlier literature two definitions of Hamiltonian were used. When the equation of motion was written in the form

$$x_{i+1} = g_i\left(x_i, u_i\right)$$

as in Katz (1962), Fan & Wang (1964), Jackson & Horn (1965), Denn & Aris (1965), Propoi (1965) then the Hamiltonian was defined as

$$H_i(x_i,u_i) = r_i(x_i,u_i) - p_{i+1} \, g_i(x_i,u_i)$$

The definition of § 214 is the straight generalization of the above definition.

On other papers by Halkin (1964), (1966), Holtzman (1966), Canon et al. (1970) the equation of motion was defined as

$$x_{i+1} = x_i + f_i(x_i,u_i)$$

and the Hamiltonian as

$$H_i(x_i,u_i) = r_i(x_i,u_i) - p_{i+1} \, f_i(x_i,u_i)$$

The appropriate generalization of the above definition to the case of the nonlinear Hamiltonian would be

$$H_i(x_i,u_i) = r_i(x_i,u_i) - \pi_{i+1}\Big[x_i + f_i(x_i,u_i)\Big] + \pi_{i+1}(x_i)$$

See the discussion in § 223a. We prefer definition 214 which is more straightforward and leads to simpler formulae.

215. If we know x_i^* and x_{i+1}^* and we want to find u_i^* we obviously have a problem of type Ⓐ. Therefore the analysis of Ⓐ in chapter 2 is directly applicable.

Specifically we get:

216. <u>Theorem</u>. Under assumption of differentiability

$$\frac{\partial \pi_i^*(x_i^*)}{\partial x_i} = \frac{\partial ub_i(x_i^*)}{\partial x_i} = \frac{\partial UB_i(x_i^*)}{\partial x_i}$$

$$i = 1,2,\cdots,N$$

Proof. The left equality is just the same as in Theorem 111. To prove the right equality we observe that

$$UB_i(x_i^*) = ub_i(x_i^*) + R_{i-1}^*$$

$$UB_i(x_i) \geq ub_i(x_i) + R_{i-1}^* \qquad \text{for all} \quad x_i \in X_i(x_{i-1}^*)$$

and the function

$$U(x_i) = UB_i(x_i) - ub_i(x_i)$$

has minimum on $X_i(x_{i-1}^*)$ at x_i^*. Then we have

$$\frac{\partial U(x_i^*)}{\partial x_i} = \frac{\partial UB_i(x_i)}{\partial x_i} - \frac{\partial ub_i(x_i)}{\partial x_i} = 0$$

which leads to the right inequality.

217. Theorem. If $\tilde{u}_i$ maximizes $H_i(x_i^*, u_i, \pi_{i+1})$ for a given function π_{i+1} over U_i then the corresponding $\tilde{x}_{i+1}$ is situated at $\{ub_{i+1}(\hat{x}_i^*)\}$.

Proof. See proof of Theorem 115 in chapter 2.

218. Theorem. $\pi_{i+1}^*(x_{i+1})$ is a support at the point $x_{i+1}^* = x_i^* + f_i(x_i^*, u_i^*)$ if and only if u_i^* maximizes $H_i(x_i^*, u_i, \pi_{i+1}^*)$ over U_i.

Proof. See proof of Theorem 116 and Corollary 116a.

218a. Theorem. $\pi_{i+1}^*(x_{i+1})$ is a strong support at the point x_{i+1}^* if and only if any solution u_i^* to:
max $H_i(x_i^*, u_i, \pi_{i+1}^*)$ over U_i satisfies
$x_i^* + f_i(x_i^*, u_i^*) = x_{i+1}^*$. Moreover any sequence u_i^*,
$i = 0, 1, \cdots, N-1$ satisfying above condition is an optimal solution to the problem Ⓑ.

Proof. See proof of Theorem 117 and Corollary 117a.

219. **Theorem.** Let $\pi_{i+1}(x_{i+1})$ be a strong support to $ub_{i+1}(x_i^*, x_{i+1})$ at any point $x_{i+1}^0 = x_i^* + f_i(x_i^*, u_i^0)$, $u_i^0 \in U_i$. Let the number k_{i+1} from the definition of a support be constant for all supports $\pi_{i+1}(x_{i+1})$. Then $\pi_{i+1}^*(x_{i+1})$ is a strong support to $ub_{i+1}(x_i^*, x_{i+1})$ at a point $x_{i+1}^* = x_i^* + f_i(x_i^*, u_i^*)$ and u_i^*, $i = 0, 1, \cdots, N$, is an optimal solution to the problem Ⓑ if and only if (u_i^*, π_{i+1}^*) is a saddle-point to the function

$$\varphi_i(u_i, \pi_{i+1}) = H_i(x_i^*, u_i, \pi_{i+1}^*) + \pi_{i+1}(x_{i+1}^*)$$

restricted to the class of strong supports with constant numbers k_{i+1}.

Proof. See proof of Theorem 121.

219a. In the case of the linear support we have

$$H_i(x_i^*, u_i, p_{i+1}) = r_i(x_i^*, u_i) - p_{i+1}^T [x_i^* + f_i(x_i^*, u_i)]$$

where p_{i+1} is a n-dimensional vector and superscript T denotes a transpose.

220. **Theorem.** A sufficient condition for the existence of a linear support to $ub_{i+1}(x_i^*, x_{i+1})$ at any point $x_{i+1}^0 \in X_{i+1}(x_i^*)$ is that $ub_{i+1}(x_i^*, x_{i+1})$ be finite and concave on a convex set $X_{i+1}(x_i^*)$ and $x_{i+1}^0 \in \text{Int } X_{i+1}(x_i^*)$.

Proof. See proof of Theorem 123.

220a. Let z be any vector in R^{n+1}. A set $S \subset R^{n+1}$ is said to be **z-directionally convex** if for each $a, b \in S$ and each $\mu \in [0, 1]$ there exists a $\beta \geq 0$ such that

$$\mu a + (1 - \mu) b + \beta z \in S$$

220b. It is seen that any convex set is z-directionally convex for any vector z since in the case of a convex set the relation of § 220a is always satisfied with $\beta = 0$.

221. <u>Theorem</u>. $ub_{i+1}(\hat{x}_i^*, x_{i+1})$ is concave if and only if $\hat{X}_{i+1}(\hat{x}_i^*)$ is e_1-directionally convex, where e_1 is the vector of the first axis in R^{n+1} , that is $e_1 = (1,0,\cdots,0)$.

<u>Proof</u>. Assume that $ub_{i+1}(x_i^*, x_{i+1})$ is concave. Then from the definition of a concave function (§ 122b) the set $X_{i+1}(x_i^*)$ is convex then for any two points $x_1, x_2 \in X_{i+1}(x_i^*)$ and for any $\mu \in [0,1]$ there exists $\beta_1 \geq 0$ that

$$ub_{i+1}\left[x_i^*, \mu x_1 + (1-\mu)x_2\right] = \mu ub_{i+1}(x_i^*, x_1) + (1-\mu)ub_{i+1}(x_i^*, x_2) + \beta_1$$

Let us take any two points $\hat{x}_1, \hat{x}_2 \in \hat{X}_{i+1}(\hat{x}_i^*)$

$$\hat{x}_1 = \begin{bmatrix} r_1 \\ x_1 \end{bmatrix} \qquad \hat{x}_2 = \begin{bmatrix} r_2 \\ x_2 \end{bmatrix} \qquad x_1, x_2 \in X_{i+1}(x_i^*)$$

For any $\mu \in [0,1]$ the point $\mu x_1 + (1-\mu)x_2 \in X_{i+1}(x_i^*)$. Let us consider the value

$$ub_{i+1}\left[x_i^*, \mu x_1 + (1-\mu)x_2\right] = \mu ub_{i+1}(x_i^*, x_1) +$$

$$+ (1-\mu)ub_{i+1}(x_i^*, x_2) + \beta_1 = \mu r_1 + (1-\mu)r_2 +$$

$$+ \mu\left[ub_{i+1}(x_i^*, x_1) - r_1\right] + (1-\mu)\left[ub_{i+1}(x_i^*, x_2) - r_2\right] + \beta_1$$

As we have

$$\left\{\mu x_1 + (1-\mu)x_2\right\} \in X_{i+1}(x_i^*)$$

$$\left\{ub_{i+1}\left[x_i^*, \mu x_1 + (1-\mu)x_2\right]\right\} \in \hat{X}_{i+1}(\hat{x}_i^*)$$

$$ub_{i+1}(x_i^*, x_j) \geq r_j \qquad j = 1,2$$

then for

$$\beta = \mu\left[ub_{i+1}\left(x_i^*, x_1\right) - r_1\right] + (1-\mu)\left[ub_{i+1}\left(x_i^*, x_2\right) - r_2\right] + \beta_1 \geq 0$$

there is

$$\mu\hat{x}_1 + (1-\mu)\hat{x}_2 + \beta e_1 \in \hat{X}_{i+1}\left(\hat{x}_i^*\right)$$

that is the set $\hat{X}_{i+1}\left(\hat{x}_i^*\right)$ is e_1-directionally convex.

Assume now that the set $\hat{X}_{i+1}\left(\hat{x}_i^*\right)$ is e_1-directionally convex, that is for each $\hat{x}_1, \hat{x}_2 \in \hat{X}_{i+1}\left(\hat{x}_i^*\right)$ and each $\mu \in [0,1]$ there exist $\beta \geq 0$ such that

$$\mu\hat{x}_1 + (1-\mu)\hat{x}_2 + \beta e_1 \in \hat{X}_{i+1}\left(\hat{x}_i^*\right)$$

We shall show first that the set $X_{i+1}\left(x_i^*\right)$ is convex (the argument is taken from Canon et al. (1970), Theorem 4.2.3, p. 84). Let

$$\hat{x}_1 = \begin{bmatrix} r_1 \\ x_1 \end{bmatrix} \qquad \hat{x}_2 = \begin{bmatrix} r_2 \\ x_2 \end{bmatrix} \qquad \begin{array}{l} \hat{x}_1, \hat{x}_2 \in \hat{X}_{i+1}\left(\hat{x}_i^*\right) \\[2mm] x_1, x_2 \in X_{i+1}\left(x_i^*\right) \end{array}$$

Then from the directional convexity for any $\mu \in [0,1]$ there exist $\beta \geq 0$ that

$$\begin{bmatrix} \mu r_1 + (1-\mu)r_2 + \beta \\ \mu x_1 + (1-\mu)x_2 \end{bmatrix} \in \hat{X}_{i+1}\left(\hat{x}_i^*\right)$$

from which we conclude that

$$\mu x_1 + (1-\mu)x_2 \in X_{i+1}\left(x_i^*\right) \qquad \mu \in [0,1]$$

that is, that $X_{i+1}\left(x_i^*\right)$ is convex.

Let us now take

$$\hat{x}_1 = \begin{bmatrix} ub_{i+1}(x_i^*, x_1) \\ \\ x_1 \end{bmatrix} \qquad \hat{x}_2 = \begin{bmatrix} ub_{i+1}(x_i^*, x_2) \\ \\ x_2 \end{bmatrix} \qquad x_1, x_2 \in X_{i+1}(x_i^*)$$

from the directional convexity for any $\mu \in [0,1]$ there exist $\beta \geq 0$ that

$$\begin{bmatrix} \mu ub_{i+1}(x_i^*, x_1) + (1-\mu) ub_{i+1}(x_i^*, x_2) + \beta \\ \\ \mu x_1 + (1-\mu) x_2 \end{bmatrix} \in \hat{X}_{i+1}(x_i^*)$$

But from the definition of $\hat{X}_{i+1}(x_i^*)$

$$\mu ub_{i+1}(x_i^*, x_1) + (1-\mu) ub_{i+1}(x_i^*, x_2) + \beta \leq$$
$$ub_{i+1}[x_i^*, \mu x_1 + (1-\mu) x_2]$$

and, as $\beta \geq 0$, we conclude that the function $ub_{i+1}(x_i^*, x_{i+1})$ is concave.

221a. In the case of a quadratic support

$$H_i(x_i^*, u_i, P_{i+1}, Q_{i+1}) = r_i(x_i^*, u_i) - P_{i+1}^T [x_i^* + f_i(x_i^*, u_i)]$$
$$- \tfrac{1}{2}[x_i^* + f_i(x_i^*, u_i)] Q_{i+1} [x_i^* + f_i(x_i^*, u_i)]$$

where the superscript T denotes a transpose, p_{i+1} is a n-dimensional vector and Q_{i+1} is a n×n symmetric matrix.

221b. A sufficient condition for the existence of a quadratic support to $ub_{i+1}(x_i^*, x_{i+1})$ at any point $x_{i+1}^o \in X_{i+1}(x_i^*)$ is that $ub_{i+1}(x_i^*, x_{i+1})$ be twice continuously differentiable in $X_{i+1}(x_i^*)$ and the function of the Hessian matrix, defined in Theorem 125

$$\frac{\partial^2 ub_{i+1}(x_i^*, x_{i+1})}{\partial x_{i+1}}$$

be bounded in $X_{i+1}(x_i^*)$, that is

$$\max_{z^T z = 1} z^T \frac{\partial^2 \, ub_{i+1}(x_i^*, x_{i+1})}{\partial x_{i+1}} z \leq M \quad \text{for all} \quad x_{i+1} \in X_{i+1}(x_i^*)$$

$\quad$ M - real number

z - real n-dimensional vector.

<u>Proof</u>. See proof of Theorem 125.

222. <u>Theorem</u>. If Ⓑ has a solution u_{i-1}^*, x_i^*, i = 1,2,...,N, then there exist supports π_{i+1}^* , i = 0,1,$\cdots$,N-1 so that

(i) $\quad u_i^*$ maximizes $H_i(x_i^*, u_i, \pi_{i+1}^*)$ over U_i ,

(ii) $\quad H_i(x_i^*, u_i, \pi_{i+1}^*) + \pi_{i+1}(x_{i+1}^*)$ has the saddle-point (u_i^*, π_{i+1}^*) ,

(iii) if the following assumptions are satisfied:

$\quad$ - $r_i(x_i, u_i^*)$ and $f_i(x_i, u_i^*)$, i = 0,1,$\cdots$,N-1 are differentiable with respect to x_i at the point x_i^* ,

$\quad$ - $UB_i(x_i)$, i = 1,2,$\cdots$,N are continuously differentiable in a small open neighbourhood of x_i^* in X_i ,

$\quad$ then

$$\frac{\partial H_i(x_i^*, u_i^*, \pi_{i+1}^*)}{\partial x_i} = - \frac{\partial \pi_i^*(x_i^*)}{\partial x_i}$$

This is the <u>nonlinear Maximum Principle</u>.

<u>Proof</u>. (i) From the assumptions and Theorem 213 there exist supports $\pi_{i+1}^*(x_{i+1})$, i = 0,1,$\cdots$,N-1 , to $ub_{i+1}(x_i^*, x_{i+1})$ at the points x_{i+1}^* . Then (i) follows from Theorem 218.

(ii)   Let the numbers k_{i+1} from the definition of supports be constant for all supports $\pi_{i+1}(x_{i+1})$. Moreover let us restrict to the class of strong supports. Then (ii) follows from Theorem 219.

(iii)   From Theorem 209 for an optimal control we have

$$UB_{i+1}(x_{i+1}^*) = UB_i(x_i^*) + r_i(x_i^*, u_i^*)$$

$$x_{i+1}^* = x_i^* + f_i(x_i^*, u_i^*)$$

Let us construct a set $\hat{\tilde{X}}_{i+1}$ given by

$$\hat{\tilde{X}}_{i+1} = \{\hat{\tilde{x}}_{i+1} : \hat{\tilde{x}}_{i+1} = \begin{bmatrix} UB_i(x_i) + r_i(x_i, u_i^*) \\ x_i + f_i(x_i, u_i^*) \end{bmatrix}, x_i \in X_i\}$$

and let $\tilde{UB}_{i+1}(x_{i+1})$ be the upper boundary of $\hat{\tilde{X}}_{i+1}$. From the construction we have

$$UB_{i+1}(x_{i+1}^*) = \tilde{UB}_{i+1}(x_{i+1}^*)$$
$$UB_{i+1}(x_{i+1}) \geq \tilde{UB}_{i+1}(x_{i+1}) \qquad x_{i+1} \in C_{i+1}$$

$$C_{i+1} = \{x_{i+1} : x_{i+1} = x_i + f_i(x_i, u_i^*), x_i \in X_i\}$$

Now, from the assumptions there exist a support $\pi_{i+1}^*(x_{i+1})$ to $UB_{i+1}(x_{i+1})$ at x_{i+1}^*, i.e. the following relations are satisfied

$$\pi_{i+1}^*(x_{i+1}^*) + k_{i+1} = UB_{i+1}(x_{i+1}^*)$$

$$\pi_{i+1}^*(x_{i+1}) + k_{i+1} \geq UB_{i+1}(x_{i+1}) \qquad x_{i+1} \in X_{i+1}$$

Comparing this and previous expressions we conclude that $\pi_{i+1}^*(x_{i+1})$ is a support to $\tilde{UB}_{i+1}(x_{i+1})$ over C_{i+1} at the point x_{i+1}^*. Then

$$\pi^*_{i+1}\left(x^*_{i+1}\right) + k_{i+1} = \tilde{UB}_{i+1}\left(x^*_{i+1}\right) = UB_i\left(x^*_i\right) + r_i\left(x^*_i, u^*_i\right)$$

$$\pi^*_{i+1}\left(x_{i+1}\right) + k_{i+1} \geq \tilde{UB}_{i+1}\left(x_{i+1}\right) \geq UB_i\left(x_i\right) + r_i\left(x_i, u^*_i\right)$$

$$x_{i+1} \in C_{i+1}, \quad x_i \in X_i$$

which we can write

$$UB_i\left(x^*_i\right) + r_i\left(x^*_i, u^*_i\right) - \pi^*_{i+1}\left[x^*_i + f_i\left(x^*_i, u^*_i\right)\right] = k_{i+1}$$

$$UB_i\left(x_i\right) + r_i\left(x_i, u^*_i\right) - \pi^*_{i+1}\left[x_i + f_i\left(x_i, u^*_i\right)\right] \leq k_{i+1} \quad x_i \in X_i$$

From the above expressions we conclude that x^*_i maximizes the function

$$P\left(x_i\right) = UB_i\left(x_i\right) + r_i\left(x_i, u^*_i\right) - \pi^*_{i+1}\left[x_i + f_i\left(x_i, u^*_i\right)\right]$$

over X_i. Now, from the assumptions $f_i\left(x_i, u^*_i\right)$, $r_i\left(x_i, u^*_i\right)$ and $UB_i\left(x_i\right)$ are differentiable at the point x^*_i. Moreover $UB_{i+1}\left(x_{i+1}\right)$ is also continuously differentiable in a small open neighbourhood of x^*_{i+1} so that a continuously differentiable support $\pi^*_{i+1}\left(x_{i+1}\right)$ exists. Then $P\left(x_i\right)$ is differentiable and

$$\frac{\partial P\left(x^*_i\right)}{\partial x_i} = \frac{\partial UB_i\left(x^*_i\right)}{\partial x_i} + \frac{\partial r_i\left(x^*_i, u^*_i\right)}{\partial x_i} - \frac{\partial \pi^*_{i+1}\left(x^*_{i+1}\right)}{\partial x_{i+1}}\left[I + \frac{\partial f_i\left(x^*_i, u^*_i\right)}{\partial x_i}\right] = 0$$

But from Theorem 216

$$\frac{\partial UB_i\left(x^*_i\right)}{\partial x_i} = \frac{\partial \pi_i\left(x^*_i\right)}{\partial x_i}$$

which after insertion to the previous formula and after rearranging terms leads to

$$\frac{\partial}{\partial x_i}\left\{r_i\left(x_i, u^*_i\right) - \pi^*_{i+1}\left[f_i\left(x_i, u^*_i\right)\right]\right\}\Bigg|_{x_i = x^*_i} = -\frac{\partial \pi^*_i\left(x^*_i\right)}{\partial x_i}$$

Now from the definition of Hamiltonian we get (iii).

222a. We summarize now the assumptions concerning supports which
we have made during proof of Theorem 222. These are:

(i)   no additional assumption,

(ii)   the numbers k_{i+1} from the definition of supports
       constant for all supports $\pi_{k+1}(x_{i+1})$ and only the
       class of strong supports has been considered,

(iii) $\pi_i^*(x_i)$ continuously differentiable in a small open
       neighbourhood of x_i^* , $i = 1,2,\cdots,N$.

222b. <u>Corollary</u>. Assume that $r_i(x_i,u_i^*)$ and $f_i(x_i,u_i^*)$,
$i = 0,1,\cdots,N-1$, are differentiable with respect to x_i ;
assume that $UB_i(x_i)$ is continuously differentiable in a
small open neighbourhood of x_i^* in X_i ; then

$$\frac{\partial r_i(x_i^*,u_i^*)}{\partial x_i} - \frac{\partial UB_{i+1}(x_{i+1}^*)}{\partial x_{i+1}} \frac{\partial f_i(x_i^*,u_i^*)}{\partial x_i} = \frac{\partial UB_{i+1}(x_{i+1}^*)}{\partial x_{i+1}} - \frac{\partial UB_i(x_i^*)}{\partial x_i}$$

<u>Proof</u>. The proof is a simple consequence of Theorem
222(iii), the definition of the Hamiltonian § 214, and
Theorem 216.

222c. We give an example showing that the assumption $x_i^* \in \text{Int } X_i$
cannot be dropped. Let us consider a 2-stage problem with

$$r_i(x_i,u_i) = u_i^2 \qquad x_0 = 0$$
$$f_i(x_i,u_i) = x_i + u_i \qquad u_i \in [0,1] \qquad i = 0,1$$

that is

$$\max (u_0^2 + u_1^2)$$
$$x_1 = u_0$$
$$x_2 = 2x_1 + u_1 = 2u_0 + u_1 \qquad u_0,u_1 \in [0,1]$$
$$\bar{x}_2 = ?$$

We have the sets of reachable states

$$x_1 \in [0,1] \qquad x_2 \in [0,3]$$

the optimal control

$$u_0^* = \begin{cases} 0 & \bar{x}_2 \leq 1 \\ \dfrac{\bar{x}_2 - 1}{2} & \bar{x}_2 > 1 \end{cases} \qquad u_1^* = \begin{cases} \bar{x}_2 & \bar{x}_2 \leq 1 \\ 1 & \bar{x}_2 > 1 \end{cases}$$

and the upper boundaries

$$UB_1(x_1) = x_1^2 \qquad UB_2(x_2) = \begin{cases} x_2^2 & 0 \leq x_2 \leq 1 \\ 1 + \frac{1}{4}(x_2 - 1)^2 & 1 < x_2 \leq 3 \end{cases}$$

Let us take $\bar{x}_2 = 0.5$. Then $u_0^* = 0$, $u_1^* = 0.5$, $x_1^* = 0$. We have

$$\frac{\partial r_i(x_i, u_i)}{\partial x_i} = 0 \qquad \frac{\partial f_i(x_i, u_i)}{\partial x_i} = 1 \qquad i = 0,1$$

and

$$\frac{\partial UB_1(0)}{\partial x_i} = 0 \qquad \frac{\partial UB_2(0.5)}{\partial x_{i+1}} = 1$$

Now the left hand side of Theorem 222b equation for $i = 1$ is equal to -1 and the right hand side to $+1$.

223. The usual linear Maximum Principle is:

<u>Theorem</u>. If linear supports exists then under assumptions of differentiability and assumptions of Theorem 222

(i) u_i^* maximizes $H_i\left(x_i^*, u_i, p_{i+1}^*\right)$ over U_i ,

(ii) $\dfrac{\partial H_i\left(x_i^*, u_i^*, p_{i+1}^*\right)}{\partial p_{i+1}^T} = \qquad -\, x_{i+1}^*$,

(iii) $\dfrac{\partial H_i\left(x_i^*, u_i^*, p_{i+1}^*\right)}{\partial x_i} = \qquad -\, p_i^*$

(iv) $H_i\left(x_i^*, u_i, p_{i+1}^*\right) + p_{i+1}^T\, x_{i+1}^*$ has the saddle-point $\left(u_i^*, p_{i+1}^*\right)$.

<u>Proof</u>. We easily recognize (i) to be a special case of Theorem 222(i), (iv) to be a special case of Theorem 222(ii) and (iii) to be a special case of Theorem 222(iii). Lastly, (ii) can be obtained by differentiating the Hamiltonian, § 219a. As we have

$$x_i^* + f_i\left(x_i^*, u_i^*\right) = x_{i+1}^*$$

then

$$H_i\left(x_i^*, u_i^*, p_{i+1}\right) = r_i\left(x_i^*, u_i^*\right) - p_{i+1}^T\, x_{i+1}^*$$

Now, differentiating both sides with respect to p_{i+1}^T we get (ii).

223b. In § 214 we have discussed two possible generalizations of two different definitions of the linear Hamiltonians. Theorem 222 is a result of the definition of Hamiltonian in § 214 which is a nonlinear generalization of the linear Hamiltonian discussed as a first in § 214a. Using a generalization of the second definition of Hamiltonian we would arrive to another version of Theorem 222 where the result 222(iii) would have the form

$$\frac{\partial H_i\left(x_i^*, u_i^*, \pi_{i+1}^*\right)}{\partial x_i} = \frac{\partial \pi_{i+1}^*\left(x_{i+1}^*\right)}{\partial x_{i+1}} - \frac{\partial \pi_i^*\left(x_i^*\right)}{\partial x_i}$$

and in the linear version in Theorem 223

(ii) $$\frac{\partial H_i\left(x_i^*, u_i^*, p_{i+1}^*\right)}{\partial p_{i+1}^T} = x_i^* - x_{i+1}^*$$

(iii) $$\frac{\partial H_i\left(x_i^*, u_i^*, p_{i+1}^*\right)}{\partial x_i} = p_{i+1}^* - p_i^*$$

This is also the classical result obtained for the second definition of the linear Hamiltonian.

224.   We now interpret $\dfrac{\partial \pi_i}{\partial x_i}$ in terms of Ⓐ . In the formulation of the problem Ⓑ we redefine:

$$x_{i+1} - x_i = f_i\left(x_i, u_i\right) - \varepsilon_i$$

with other conditions unchanged. We define $UB_N\left(\bar{x}_N, \varepsilon_0, \varepsilon_1, \cdots, \varepsilon_{N-1}\right) = UB_N\left(\bar{x}_N, \varepsilon\right)$ as the optimal value for given ε and we assume that the conditions of § 208 are satisfied for the redefined problem.

225.   Theorem.   Assume that for optimal sequences u^*_{i-1}, x^*_i, $i = 1, 2, \ldots, N$ the following conditions are satisfied:

(i) $ub_{i+1}\left(x^*_i, x_{i+1}\right)$ is continuously differentiable in an open neighbourhood of x^*_{i+1} in X_{i+1}, $i = 1, 2, \cdots, N$,

(ii) $\pi^*_{i+1}\left(x_{i+1}\right)$ — a support to $ub_{i+1}\left(x^*_i, x_{i+1}\right)$ at the point x^*_{i+1} — is differentiable at x^*_{i+1}, $i = 1, 2, \cdots, N$,

(iii) $UB_N\left(\bar{x}_N, \epsilon\right)$ is continuously differentiable in an open neighbourhood of $\epsilon = 0$,

(iv) $\dim u_i = m = n = \dim x_i$,

(v) $f_i\left(x^*_i, u_i\right)$ is continuously differentiable with respect to u_i in an open neighbourhood of u^*_i in U_i and $\dfrac{\partial f_i\left(x^*_i, u_i\right)}{\partial u_i}$ is nonsingular, $i = 0, 1, \cdots, N$,

then

$$\frac{\partial \pi^*_{i+1}\left(x^*_{i+1}\right)}{\partial x_{i+1}} = \frac{\partial UB_N\left(\bar{x}_N, 0\right)}{\partial \epsilon_i}$$

Proof.   Let x^*_i, x^*_{i+1} be optimal states.   Let us consider two equations

$$x^*_{i+1} = x^*_i + f_i\left(x^*_i, u^*_i\right)$$

$$x^*_{i+1} = x^*_i + f_i\left(x^*_i, u^\epsilon_i\right) - \epsilon_i$$

We claim that there exists a small open neighbourhood of $\varepsilon_i = 0$ that u_i^ε is an internal point of U_i and x_{i+1}^* is an internal point of $X_{i+1}^\varepsilon(x_i^*)$, the set of states reachable from x_i^* for the problem of § 224. We give a short sketch of a proof. From assumptions (iv) and (v), for fixed x_i^*, the function f_i is continuous in an open neighbourhood of u_i^*, say $D_i \subset U_i$, and has a continuous inverse, such that $x_{i+1}^* - x_i^*$ is the only point that $f_i^{-1}(x_{i+1}^* - x_i^*) = u_i^*$. Then both f_i and f_i^{-1} map open sets to open sets. Consider now the second equation. For fixed x_i^* and x_{i+1}^* let E_i be the image of D_i. Then for any $\varepsilon_i \in E_i$ there exists $u_i^\varepsilon \in D_i \subset U_i$ which satisfies second equation, and D_i is an open set. This proves the first part of our claim. Furthermore for any fixed $\varepsilon_i \in E_i$ and fixed x_i^* the image of D_i is an open neighbourhood of x_{i+1}^*, say C_{i+1}^ε. Because $D_i \subset U_i$ then $C_{i+1}^\varepsilon \subset X_{i+1}^\varepsilon(x_i^*)$ which proves the second part of our claim. As f_i^{-1} is a continuous function we have moreover the result $u_i^\varepsilon \xrightarrow[\varepsilon \to 0]{} u_i^*$.

Now, let $\{\varepsilon_i^k\}_{k=1}^\infty$ be a sequence that $\varepsilon_i^k \in E_i$, $k = 1,2,\cdots$ and $\lim_{n \to \infty} \varepsilon_i^k = 0$, and $\{u_i^k\}_{k=1}^\infty$ be a sequence that

$$x_{i+1}^* = x_i^* + f_i(x_i^*, u_i^k) - \varepsilon_i^k \qquad u_i^i \neq u_i^* \qquad k = 1,2,\cdots$$

and $\hat{x}_{i+1}^* = \begin{bmatrix} r_{i+1}(x_{i+1}^*, u_i^k) \\ x_{i+1}^* \end{bmatrix}$ is situated at the upper boundary of $X_{i+1}^\varepsilon(x_i^*)$. Then $u_i^k \in D_i$. Now, from assumption (v)

$$\varepsilon_i^k = -x_{i+1}^* + x_i^* + f_i(x_i^*, u_i^*) + \frac{\partial f_i(x_i^*, u_i^* + \theta_u^k \Delta u_i^k)}{\partial u_i}(u_i^k - u_i^*) =$$

$$= \frac{\partial f_i(x_i^*, u_i^* + \theta_u^k \Delta u_i^k)}{\partial u_i}(u_i^k - u_i^*) \qquad \begin{aligned} \Delta u_i^k &= u_i^k - u_i^* \\ 0 &\leq \theta_u^k \leq 1 \end{aligned}$$

Defining $UB_N(\bar{x}_N, 0, \cdots, 0, \epsilon_i^k, 0, \cdots, 0) = UB_N(\bar{x}_N, \epsilon_i^k)$ we have from assumption (iii)

$$UB_N(\bar{x}_N, \epsilon_i^k) - UB_N(\bar{x}_N, 0) =$$

$$= \frac{\partial UB_N(\bar{x}_N, \theta_\epsilon^k \epsilon_i^k)}{\partial \epsilon_i} \frac{\partial f_i(x_i^*, u_i^* + \theta_u^k \Delta u_i^k)}{\partial u_i}(u_i^k - u_i^*) \qquad k = 1, 2, \cdots$$

$$0 \leq \theta_\epsilon^k \leq 1$$

From the other side we have

$$UB_N(\bar{x}_N, 0) = ub_{i+1}(x_i^*, x_{i+1}^*) + \sum_{j=i+1}^{N-1} r_j(x_j^*, u_j^*)$$

$$UB_N(\bar{x}_N, \epsilon_i^k) = ub_{i+1}^\epsilon(x_i^*, x_{i+1}^*) + \sum_{j=i+1}^{N-1} r_j(x_j^*, u_j^*)$$

where $ub_{i+1}^\epsilon(x_i^*, x_{i+1})$ is an upper boundary of the set $X_{i+1}^\epsilon(x_i^*)$. Then

$$UB_N(\bar{x}_N, \epsilon_i^k) - UB(\bar{x}_N, 0) = ub_{i+1}^\epsilon(x_i^*, x_{i+1}^*) - ub_{i+1}(x_i^*, x_{i+1}^*)$$

But

$$ub_{i+1}^\epsilon(x_i^*, x_{i+1}^*) = ub_i(x_{i-1}^*, x_i^*) + r_i(x_i^*, u_i^k)$$

For given u_i^k we have now

$$x_{i+1}^k = x_i^* + f_i(x_i^*, u_i^k)$$

As $u_i^k \in D_i \subset U_i$ then $x_{i+1}^k \in X_{i+1}(x_i^*)$. So we conclude that

$$ub_{i+1}^\epsilon(x_i^*, x_{i+1}^*) = ub_{i+1}(x_i^*, x_{i+1}^k)$$

and from assumption (i)

$$UB_N(\bar{x}_N, \varepsilon_i^k) - UB(\bar{x}_N, 0) = ub_{i+1}(x_i^*, x_{i+1}^k) - ub_{i+1}(x_i^*, x_{i+1}^*) =$$

$$= \frac{\partial ub_{i+1}(x_i^*, x_{i+1}^* + \theta_x^k \Delta x_{i+1}^k)}{\partial x_{i+1}}(x_{i+1}^k - x_{i+1}^*) =$$

$$= \frac{\partial ub_{i+1}(x_i^*, x_{i+1}^* + \theta_x^k \Delta x_{i+1}^k)}{\partial x_{i+1}} \frac{\partial f_i(x_i^*, u_i^* + \theta_u^k \Delta u_i^k)}{\partial u_i}(u_i^k - u_i^*)$$

$$k = 1, 2, \cdots$$

$$\Delta x_{i+1}^k = x_{i+1}^k - x_{i+1}^* \qquad \Delta u_i^k = u_i^k - u_i^* \qquad 0 \le \theta_x^k \le 1$$

Now we have

$$\frac{\partial UB_N(\bar{x}_N, \theta_\varepsilon^k \varepsilon_i^k)}{\partial \varepsilon_i} \frac{\partial f_i(x_i^*, u_i^* + \theta_u^k \Delta u_i^k)}{\partial u_i}(u_i^k - u_i^*) =$$

$$= \frac{\partial ub_{i+1}(x_i^*, x_{i+1}^* + \theta_x^k \Delta x_{i+1}^k)}{\partial x_{i+1}} \frac{\partial f_i(x_i^*, u_i^* + \theta_u^k \Delta u_i^k)}{\partial u_i}(u_i^k - u_i^*)$$

$$k = 1, 2, \cdots$$

that is

$$\left[\frac{\partial UB_N(\bar{x}_N, \theta_\varepsilon^k \varepsilon_i^k)}{\partial \varepsilon_i} - \frac{\partial ub_{i+1}(x_i^*, x_{i+1}^* + \theta_x^k \Delta x_{i+1}^k)}{\partial x_{i+1}}\right] \frac{\partial f_i(x_i^*, u_i^* + \theta_u^k \Delta u_i^k)}{\partial u_i}$$

$$(u_i^k - u_i^*) = 0$$

As $u_i^k - u_i^* \ne 0$ and $\dfrac{\partial f_i(x_i^*, u_i^* + \theta_u^k \Delta u_i^k)}{\partial u_i}$ is nonsingular, then

$$\frac{\partial UB_N(\bar{x}_N, \theta_\varepsilon^k \varepsilon_i^k)}{\partial \varepsilon_i} = \frac{\partial ub_{i+1}(x_i^*, x_{i+1}^* + \theta_x^k \Delta x_{i+1}^k)}{\partial x_{i+1}} \qquad k = 1, 2, \cdots$$

Now, letting $\varepsilon_i^k \to 0$ we have $\Delta x_{i+1}^k \to 0$ and

$$\frac{\partial UB_N(\bar{x}_N, 0)}{\partial \varepsilon_i} = \frac{\partial ub_{i+1}(x_i^*, x_{i+1}^*)}{\partial x_{i+1}}$$

Because of the assumptions of differentiability of $UB_N(\bar{x}_N, 0)$ and $ub_{i+1}(x_i^*, x_{i+1}^*)$ the limit for the special sequence $\{\varepsilon_i^k\}$ is equal to the limit for any sequence and equal to derivatives. Now, we complete the proof taking into account Theorem 216.

225a. We give some examples illustrating Theorem 225.

Example 1. Consider a problem

$$\max\left[-(u_0)^2 - (u_1)^2\right]$$

$$x_1 = 2x_0 + u_0 \qquad \bar{x}_0 = 0 \qquad u_0, u_1 \in R$$

$$x_2 = 2x_1 + u_1 \qquad \bar{x}_2 = \bar{x}_2$$

That is

$$\max\left[-(u_0)^2 - (u_1)^2\right]$$

$$x_1 = u_0$$

$$\bar{x}_2 = 2u_0 + u_1$$

The Lagrangian is equal to

$$L = -(u_0)^2 - (u_1)^2 + \lambda(2u_0 + u_1 - \bar{x}_2)$$

and the optimal solution is

$$u_0^* = \frac{2}{5}\bar{x}_2 \qquad u_1^* = \frac{1}{5}\bar{x}_2 \qquad x_1^* = \frac{2}{5}\bar{x}_2 \qquad x_2^* = \bar{x}_2$$

Then we have

$$UB_1(x_1) = -(x_1)^2 \qquad UB_2(x_2) = -\frac{1}{5}(x_2)^2$$

and

$$\frac{\partial UB_1(x_1^*)}{\partial x_1} = -\frac{4}{5}\bar{x}_2 \qquad \frac{\partial UB_2(\bar{x}_2)}{\partial x_2} = -\frac{2}{5}\bar{x}_2$$

Now, our redefined problem is

$$\max\left[-\left(u_0\right)^2 - \left(u_1\right)^2\right]$$

$$x_1 = 2x_0 + u_0 - \varepsilon_0 \qquad \bar{x}_0 = 0 \qquad u_0, u_1 \in R$$

$$x_2 = 2x_1 + u_1 - \varepsilon_1 \qquad \bar{x}_2 = \bar{x}_2$$

That is

$$\max\left[-\left(u_0\right)^2 - \left(u_1\right)^2\right]$$

$$x_1 = u_0 - \varepsilon_0$$

$$\bar{x}_2 = 2u_0 + u_1 - 2\varepsilon_0 - \varepsilon_1$$

The Lagrangian is equal to

$$L = -\left(u_0\right)^2 - \left(u_1\right)^2 + \lambda\left(2u_0 + u_1 - 2\varepsilon_0 - \varepsilon_1 - \bar{x}_2\right)$$

and the optimal solution is

$$u_0^* = \tfrac{2}{5}\left(\bar{x}_2 + 2\varepsilon_0 + \varepsilon_1\right) \qquad u_1^* = \tfrac{1}{5}\left(\bar{x}_2 + 2\varepsilon_0 + \varepsilon_1\right)$$

$$x_1^* = \tfrac{2}{5}\left(\bar{x}_2 + 2\varepsilon_0 + \varepsilon_1\right) \qquad x_2^* = \bar{x}_2$$

Then we have

$$UB_1\left(x_1,\varepsilon\right) = -\left(x_1 + \varepsilon_0\right)^2 \qquad UB_2\left(x_2,\varepsilon\right) = -\tfrac{1}{5}\left(x_2 + 2\varepsilon_0 + \varepsilon_1\right)^2$$

and

$$\frac{\partial UB_2\left(\bar{x}_2,0\right)}{\partial\varepsilon_0} = -\tfrac{4}{5}\bar{x} \qquad \frac{\partial UB_2\left(\bar{x}_2,0\right)}{\partial\varepsilon_1} = -\tfrac{2}{5}\bar{x}$$

which should be compared with values on the bottom of the previous page. This is an example where Theorem 225 works.

Example 2. We show now that violation of the assumption
(iv) may lead to a wrong result. Consider a problem

$$\max \left[-\left(u_0\right)^2 - \left(u_1\right)^2 \right]$$

$$\begin{cases} x_1^1 = x_0^1 + x_0^2 + u_0 \\ \\ x_1^2 = x_0^2 + u_0 \end{cases} \qquad \bar{x}_0 = \begin{bmatrix} \bar{x}_0^1 \\ \\ \bar{x}_0^2 \end{bmatrix} = 0 \qquad u_0 \in R$$

$$\begin{cases} x_2^1 = x_1^1 + x_1^2 + u_1 \\ \\ x_2^2 = x_1^2 + u_1 \end{cases} \qquad \bar{x}_2 = \begin{bmatrix} \bar{x}_2^1 \\ \\ \bar{x}_2^2 \end{bmatrix} = \begin{bmatrix} \bar{x}_2^1 \\ \\ \bar{x}_2^2 \end{bmatrix}$$

That is

$$\max \left[-\left(u_0\right)^2 - \left(u_1\right)^2 \right]$$

$$\begin{cases} x_1^1 = u_0 \\ \\ x_1^2 = u_0 \end{cases}$$

$$\begin{cases} \bar{x}_2^1 = 2u_0 + u_1 \\ \\ \bar{x}_2^2 = u_0 + u_1 \end{cases}$$

The Lagrangian is equal to

$$L = -\left(u_0\right)^2 - \left(u_1\right)^2 + \lambda_1 \left(2u_0 + u_1 - \bar{x}_2^1\right) + \lambda_2 \left(u_0 + u_1 - \bar{x}_1^2\right)$$

and the optimal solution is

$$u_0^* = \bar{x}_2^1 - \bar{x}_2^2 \qquad\qquad u_1^* = 2\bar{x}_2^2 - \bar{x}_2^1$$

$$x_1^{1*} = x_1^{2*} = \bar{x}_2^1 - \bar{x}_2^2$$

$$x_2^{1*} = \bar{x}_2^1 \qquad x_2^{2*} = \bar{x}_2^2$$

Then we have

$$UB_1(x_1) = -(x_1^1)^2 = -(x_1^2)^2 = -\frac{1}{2}(x_1^1)^2 - \frac{1}{2}(x_1^2)^2$$

$$UB_2(x_2) = -(x_2^1 - x_2^2)^2 - (2x_2^2 - x_2^1)^2$$

and

$$\left.\begin{array}{l}\dfrac{\partial UB_1(x_1^*)}{\partial x_1^1} = \\[4mm] \dfrac{\partial UB_2(x_1^*)}{\partial x_1^2} = \end{array}\right\} \text{ depends on which formula we choose}$$

The reason of this strange phenomena is that the set of extended states X_1 has the following form

$$X_1 = \left\{ \begin{bmatrix} x_1^1 \\ \\ x_1^2 \end{bmatrix} : \quad x_1^1 = x_1^2, \quad x_1^1, x_1^2 \in R \right\}$$

which is a line in a space (x_1^1, x_1^2) :

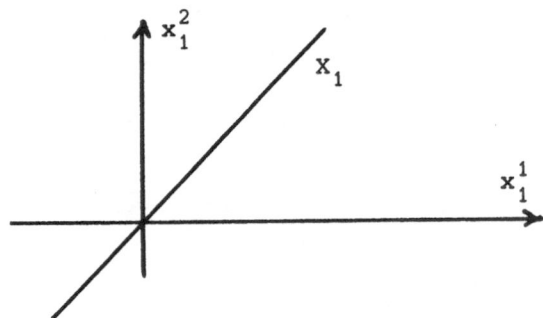

The function $UB_1(x_1)$ is defined only on the line X_1, and any differentiation with respect to x_1^1 or x_1^2 leeds out of the set X_1, which has no interior points. This is then a case where Theorem 225 does not work.

Example 3. We show now that the violation of the assumption (v) may lead to a wrong result. Consider a problem

$$\max\left[-(u_0^1)^2 - (u_0^2)^2 - (u_1^1)^2 - (u_1^2)^2\right]$$

$$\begin{cases} x_1^1 = x_0^2 + 4u_0^1 + 2u_0^2 \\[2mm] x_1^2 = x_0^1 + 8u_0^1 + 4u_0^2 \end{cases} \qquad \bar{x}_0 = \begin{bmatrix} \bar{x}_0^1 \\ \bar{x}_0^2 \end{bmatrix} = 0 \qquad u_0^1, u_0^2, u_1^1, u_1^2 \in R$$

$$\begin{cases} x_2^1 = x_1^2 + 4u_1^1 + 2u_1^2 \\[2mm] x_2^2 = x_1^1 + 8u_1^1 + 4u_1^2 \end{cases} \qquad \bar{x}_2 = \begin{bmatrix} \bar{x}_2^1 \\ \bar{x}_2^2 \end{bmatrix} = \begin{bmatrix} \bar{x}_2^1 \\ \bar{x}_2^2 \end{bmatrix}$$

That is

$$\max\left[-(u_0^1)^2 - (u_0^2)^2 - (u_1^1)^2 - (u_1^2)^2\right]$$

$$\begin{cases} x_1^1 = 4u_0^1 + 2u_0^2 \\[2mm] x_1^2 = 8u_0^1 + 4u_0^2 \end{cases}$$

$$\begin{cases} \bar{x}_2^1 = 8u_0^1 + 4u_0^2 + 4u_1^1 + 2u_1^2 \\[2mm] \bar{x}_2^2 = 4u_0^1 + 2u_0^2 + 8u_1^1 + 4u_1^2 \end{cases}$$

We have

$$f_i(x_i, u_i) = \begin{bmatrix} 4u_i^1 + 2u_i^2 \\[2mm] 8u_i^1 + 4u_i^2 \end{bmatrix}$$

and

$$\frac{\partial f_i(x_i, u_i)}{\partial u_i} = \begin{bmatrix} 4 & 2 \\ 8 & 4 \end{bmatrix} \qquad \det \begin{bmatrix} 4 & 2 \\ 8 & 4 \end{bmatrix} = 0$$

Now the set X_1 of reachable states is

$$X_1 = \left\{ \begin{bmatrix} x_1^1 \\ x_1^2 \end{bmatrix} : x_1^2 = 2x_1^1 , \quad x_1^1, x_1^2 \in R \right\}$$

which is once more a line in a space (x_1^1, x_1^2)

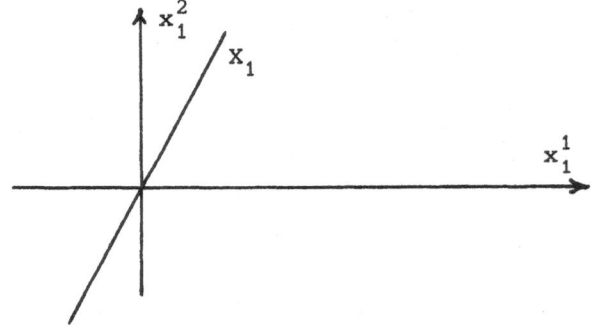

and the same phenomena like in Example 2 occurs.

226. We now return to optimality condition § 209. Given the
      greater upper boundary $\{UB_i\}$ we can generate $\{UB_{i+1}\}$
      by using the condition given below.

Theorem. Let $V_i \subset U_i \times X_i$ be a set of all pairs (u_i, x_i)
such that the control u_i transforms x_i to x_{i+1}, i.e.
$V_i = \{(u_i, x_i) : u_i \in U_i , \quad x_i \in X_i , \quad x_i + f_i(x_i, u_i) = x_{i+1}\}$.
Then

$$UB_{i+1}(x_{i+1}) = \max_{(u_i, x_i) \in V_i} \left[r_i(x_i, u_i) + UB_i(x_i) \right]$$

Proof. From the definition of the upper boundary we have

$$UB_{i+1}(x_{i+1}) = \max_{u_0 \in U_0} \sum_{j=0}^{i} r_j(x_j,u_j) = \max_{u_i \in U_i} \left[r_i(x_i,u_i) + \max_{u_0 \in U_0} \sum_{j=0}^{i-1} r_j(x_j,u_j) \right]$$

$$\vdots \qquad\qquad x_{i+1}-x_i = f_i(x_i,u_i) \qquad \vdots$$

$$u_{i-1} \in U_{i-1} \qquad\qquad\qquad u_{i-1} \in U_{i-1}$$

$$u_i \in U_i \qquad\qquad\qquad x_1-\bar{x}_0 = f_0(\bar{x}_0,u_0)$$

$$x_1-\bar{x}_0 = f_0(\bar{x}_0,u_0) \qquad\qquad \vdots$$

$$\vdots \qquad\qquad\qquad x_i-x_{i-1} = f_{i-1}(x_{i-1},u_{i-1})$$

$$x_i-x_{i-1} = f_{i-1}(x_{i-1},u_{i-1})$$

$$x_{i+1}-x_i = f_i(x_i,u_i)$$

$$= \max_{u_i \in U_i} \left[r_i(x_i,u_i) + UB_i(x_i) \right] = \max_{(u_i,x_i) \in V_i} \left[r_i(x_i,u_i) + UB_i(x_i) \right]$$

$$x_{i+1}-x_i = f_i(x_i,u_i)$$

227. The recursion in § 226 generates $\{UB_{i+1}\}$ from $\{UB_i\}$. Normally the <u>Principle of Optimality</u>, and then <u>Dynamic Programming</u>, is presented in a backward version (to be presented in § 229). Due to the direction of transfer $x_{i+1} - x_i = f_i(x_i,u_i)$, where x_i and u_i uniquely specify x_{i+1}, we shall in the backward version only maximize with respect to u_i, while in the forward version (§ 226) must treat both u_i and x_i as variables.

228. We consider the <u>reverse greater upper boundary</u> at stage i $\{RUB_i\}$ as a set, $RUB_i(x_i)$ as a function (defined for all x_i from which $\bar{x}_N$ can be reached) which we define as

$$RUB_i(x_i) = \max \sum_{j=i}^{N-1} r_j(x_j,u_j)$$

Here the optimization is with respect to $u_i, u_{i+1}, \cdots, u_{N-1}$ restricted by $u_i \in U_i, \cdots, u_{N-1} \in U_{N-1}$ and

$$x_{i+1} - x_i = f_i(x_i, u_i)$$

$$x_{i+2} - x_{i+1} = f_{i+1}(x_{i+1}, u_{i+1})$$

$$\vdots$$

$$\bar{x}_N - x_{N-1} = f_{N-1}(x_{N-1}, u_{N-1})$$

We also define $RUB_N(\bar{x}_N) = 0$.

229. __Theorem.__

$$RUB_i(x_i) = \max_{u_i \in U_i} \{r_i(x_i, u_i) + RUB_{i+1}[x_i + f_i(x_i, u_i)]\}$$

__Proof.__ The proof of this theorem is rather well known.
We repeat it, however, to make the book self-contained.
From the definition 228 we have

$$RUB_i(x_i) = \max_{\substack{u_i \in U_i \\ u_{i+1} \in U_{i+1} \\ \vdots \\ u_{N-1} \in U_{N-1}}} \sum_{j=i}^{N-1} r_j(x_j, u_j) = \max_{\substack{u_i \in U_i \\ x_{i+1} - x_i = f_i(x_i, u_i)}} \left[r_i(x_i, u_i) + \max_{\substack{u_{i+1} \in U_{i+1} \\ \vdots \\ u_{N-1} \in U_{N-1}}} \sum_{j=i+1}^{N-1} r_j(x_j, u_j) \right] =$$

$$\begin{array}{ll} x_{i+1} - x_i = f_i(x_i, u_i) & x_{i+2} - x_{i+1} = f_{i+1}(x_{i+1}, u_{i+1}) \\ x_{i+2} - x_{i+1} = f_{i+1}(x_{i+1}, u_{i+1}) & \vdots \\ \vdots & \bar{x}_N - x_{N-1} = f_{N-1}(x_{N-1}, u_{N-1}) \\ \bar{x}_N - x_{N-1} = f_{N-1}(x_{N-1}, u_{N-1}) & \end{array}$$

$$= \max_{u_i \in U_i} \left[r_i(x_i, u_i) + RUB_{i+1}(x_{i+1}) \right] =$$

$$x_{i+1} - x_i = f_i(x_i, u_i)$$

$$= \max_{u_i \in U_i} \{r_i(x_i, u_i) + RUB_{i+1}[x_i + f_i(x_i, u_i)]\}$$

231. <u>Theorem.</u>

$$-RUB_i\left(x_i^*\right) + UB_N\left(\bar{x}_N\right) = UB_i\left(x_i^*\right)$$

$$-RUB_i\left(x_i\right) + UB_N\left(\bar{x}_N\right) \geq UB_i\left(x_i\right)$$

for all x_i, for which the two functions are both defined.

<u>Proof.</u> From the assumptions 208 there exist optimal se-
quences u_{j-1}^*, x_j^*, $j = 1, 2, \cdots, N$, that

$$UB_N\left(\bar{x}_N\right) = \sum_{j=0}^{N-1} r_j\left(x_j^*, u_j^*\right) = \max_{u_0 \in U_0} \sum_{j=0}^{N-1} r_j\left(x_j, u_j\right)$$

$$x^* = \bar{x}_0$$
$$x_1^* - x_0^* = f_0\left(x_0^*, u_0^*\right)$$
$$\vdots$$
$$x_i^* - x_{i-1}^* = f_{i-1}\left(x_{i-1}^*, u_{i-1}^*\right)$$
$$x_{i+1}^* - x_i^* = f_i\left(x_i^*, u_i^*\right)$$
$$\vdots$$
$$\bar{x}_N - x_{N-1}^* = f_{N-1}\left(x_{N-1}^*, u_{N-1}^*\right)$$

$$\vdots$$
$$u_{i-1} \in U_{i-1}$$
$$u_i \in U_i$$
$$u_{i+1} \in U_{i+1}$$
$$\vdots$$
$$u_{N-1} \in U_{N-1}$$

$$x_1 - \bar{x}_0 = f_0\left(\bar{x}_0, u_0\right)$$
$$\vdots$$
$$x_i - x_{i-1} = f_{i-1}\left(x_{i-1}, u_{i-1}\right)$$
$$x_{i+1} - x_i = f_i\left(x_i, u_i\right)$$
$$\vdots$$
$$\bar{x}_N - x_{N-1} = f_{N-1}\left(x_{N-1}, u_{N-1}\right)$$

Then as x_i^* is situated at the above specified optimal
trajectory

$$UB_N\left(\bar{x}_N\right) = \max_{u_0 \in U_0} \sum_{j=0}^{i-1} r_j\left(x_j, u_j\right) + \max_{u_i \in U_i} \sum_{j=i}^{N-1} r_j\left(x_j, u_j\right) =$$

$$\vdots$$
$$u_{i-1} \in U_{i-1}$$

$$\vdots$$
$$u_{N-1} \in U_{N-1}$$

$$x_1 - \bar{x}_0 = f_0\left(x_0, u_0\right)$$
$$\vdots$$
$$x_i^* - x_{i-1} = f_{i-1}\left(x_{i-1}, u_{i-1}\right)$$

$$x_{i+1} - x_i^* = f_i\left(x_i^*, u_i\right)$$
$$\vdots$$
$$\bar{x}_N - x_{N-1} = f_i\left(x_{N-1}, u_{N-1}\right)$$

$$= UB_i\left(x_i^*\right) + RUB_i\left(x_i^*\right)$$

which proves the first expression of the Theorem.

To prove the second expression let us observe that for any u_{j-1}, x_j, $j = 1, 2, \cdots, N$ satisfying restrictions we have

$$UB_N(\bar{x}_N) = \sum_{j=0}^{N-1} r_j(x_j^*, u_j^*) \geq \sum_{j=0}^{N-1} r_j(x_j, u_j)$$

Now, repeating the earlier argumentation we conclude that

$$UB_N(\bar{x}_N) \geq UB_i(x_i) + RUB_i(x_i)$$

231a. <u>Corollary</u>. $-RUB_i(\cdot)$ is a support to $\{ub_i(\hat{x}_{i-1}^*, \cdot)\}$ for those x_i, for which $RUB_i(\cdot)$ and $ub_i(\hat{x}_{i-1}^*, \cdot)$ are both defined.

<u>Proof</u>. The corollary is a simple consequence of Theorems 231, 209, 210 and observation of § 207.

232. <u>Theorem</u>. If and only if $\hat{x}_i^*$ is optimal it is situated both at $\{UB_i\}$ and $\{-RUB_i + UB_N(\bar{x}_N)\}$.

<u>Proof</u>: It has been already proved in Theorem 231 that if $\hat{x}_i^*$ is optimal then it is situated both at $\{UB_i\}$ and $\{-RUB_i + UB_N(\bar{x}_N)\}$. Now we only prove the reverse statement. Let us assume that $\hat{x}_i^0$ is situated both at $\{UB_i\}$ and $\{-RUB_i + UB_N(\bar{x}_N)\}$. That is we have

$$UB_i(x_i^0) = -RUB_i(x_i^0) + UB_N(\bar{x}_N) \qquad (*)$$

Now, from the definitions of $UB_i(\cdot)$ and $RUB_i(\cdot)$

$$UB_i\left(x_i^0\right) = \max_{\substack{u_0 \in U_0 \\ \vdots \\ u_{i-1} \in U_{i-1}}} \sum_{j=0}^{i-1} r_j\left(x_j, u_j\right) \qquad \begin{aligned} x_1 - \bar{x}_0 &= f_0\left(\bar{x}_0, u_0\right) \\ &\vdots \\ x_i^0 - x_{i-1} &= f_{i-1}\left(x_{i-1}, u_{i-1}\right) \end{aligned}$$

$$RUB_i\left(x_i^0\right) = \max_{\substack{u_i \in U_i \\ \vdots \\ u_{N-1} \in U_{N-1}}} \sum_{j=i}^{N-1} r_j\left(x_j, u_j\right) \qquad \begin{aligned} x_{i+1} - x_i^0 &= f_i\left(x_i^0, u_i\right) \\ &\vdots \\ \bar{x}_N - x_{N-1} &= f_{N-1}\left(x_{N-1}, u_{N-1}\right) \end{aligned}$$
$$x_i = x_i^0$$

Let us define the sequences which maximize the first sum (in definition of UB_i) by $u_0^0, u_1^0, \cdots, u_{i-1}^0$; $\bar{x}_0 = x_0^0, x_1^0,$ $\cdots, x_{i-1}^0, x_i^0$ and the sequences which minimize the second sum (in definition of RUB_i) by $u_i^0, u_{i-1}^0, \cdots, u_{N-1}^0$; $x_i^0, x_{i+1}^0, \cdots, x_{N-1}^0, x_N^0 = \bar{x}_N$. Then we have

$$UB_i\left(x_i^0\right) = \sum_{j=0}^{i-1} r_j\left(x_j^0, u_j^0\right)$$

$$RUB_i\left(x_i^0\right) = \sum_{j=i}^{N-1} r_j\left(x_j^0, u_j^0\right)$$

and from the equality (*)

$$UB_N\left(\bar{x}_N\right) = UB_i\left(x_i^0\right) + RUB_i\left(x_i^0\right) = \sum_{j=0}^{N-1} r_j\left(x_j^0, u_j^0\right)$$

But this means that the sequences $u_0^0, u_1^0, \cdots, u_{N-1}^0$; $x_0^0, x_1^0, \cdots, x_{N-1}^0$ are optimal. Then x_i^0 is also optimal as a special case.

233. Theorem.

$$UB_{i+1}\left(x_{i+1}^*\right) - UB_i\left(x_i^*\right) = r_i\left(x_i^*, u_i^*\right) = -RUB_{i+1}\left(x_{i+1}^*\right) + RUB_i\left(x_i^*\right)$$

<u>Proof</u>. From Theorem 209 x^*_{i+1} is situated at $\{UB_{i+1}\}$ and x^*_i at $\{UB_i\}$. Then we have

$$UB_{i+1}\left(x^*_{i+1}\right) = \sum_{j=0}^{i} r_j\left(x^*_j,u^*_j\right) = r_i\left(x^*_i,u^*_i\right) + \sum_{j=0}^{i-1} r_j\left(x^*_j,u^*_j\right) =$$

$$= r_i\left(x^*_i,u^*_i\right) + UB_i\left(x^*_i\right)$$

which proves the left-hand equality. To prove the right-hand equality let us observe that

$$UB_{i+1}\left(x^*_{i+1}\right) - UB_i\left(x^*_i\right) = r_i\left(x^*_i,u^*_i\right)$$

and from Theorem 231

$$UB_{i+1}\left(x^*_{i+1}\right) = -RUB_{i+1}\left(x^*_{i+1}\right) + UB_N\left(\bar{x}_N\right)$$

$$UB_i\left(x^*_i\right) = -RUB_i\left(x^*_i\right) + UB_N\left(\bar{x}_N\right)$$

Inserting $UB_i\left(x^*_i\right)$ and $UB_{i+1}\left(x^*_{i+1}\right)$ from the two last expressions to the previous equality we get the required result.

234. <u>Theorem</u>. Under assumptions of differentiability

$$\frac{\partial UB_i\left(x^*_i\right)}{\partial x_i} = \frac{\partial ub\left(x^*_{i-1},x^*_i\right)}{\partial x_i} = \frac{\partial \pi_i\left(x^*_i\right)}{\partial x_i} = - \frac{\partial RUB_i\left(x^*_i\right)}{\partial x_i}$$

<u>Proof</u>. The two first equalities have already been proved in Theorem 216. The last equality is then a simple consequence of Theorem 231 under assumption of differentiability.

3.1 Literature

This chapter is based on the report

> H.F. Ravn: Upper Boundary Methods. IMSOR, Research Report No. 10, 1980,

see also

> R.V.V. Vidal: Notes in Static and Dynamic Optimization. IMSOR, 1981.

The early (incorrect) derivation of the Linear Discrete Maximum Principle is in

> S. Katz: Best Operating points for staged systems. Ind. & Eng. Chem. Fundamentals, Vol. 1, No. 4, 1962, pp. 226-240.

> L.T. Fan, Ch. S. Wang: The Discrete Maximum Principle. Wiley, 1964,

and the discussion around it

> F. Horn, R. Jackson: Discrete Maximum Principle. Ind. & Eng. Chem. Fundamentals, Vol. 4, No. 1, pp. 110-112, No. 4, pp. 487-488, 1965.

> R. Jackson, F. Horn: On Discrete Analogues of Pontryagin's Maximum Principle. Int. J. Control, Vol. 1, No. 4, 1965, pp. 389-395.

> M.M. Denn: Discrete Maximum Principle. Ind. & Eng. Chem. Fundamentals, Vol. 4, No. 2, 1965, pp. 240.

The correct derivation of the Linear Discrete Maximum Principle with convexity assumption is in

H. Halkin: Optimal Control for Systems Described by Difference Equations. In: C.T. Leonder (ed.): Advances in
Control Systems, Vol. I. Academic Press, 1964.

H. Halkin: A Maximum Principle of the Pontryagin Type for
Systems Described by Nonlinear Difference Equations.
SIAM J. Control, Vol. 4, No. 1, 1966, pp. 90-111.

A.I. Propoi: The Maximum Principle for Discrete Control
Systems. Automation and Remote Control, Vol. 26, No. 7,
1965, pp. 1167-1177,

and with directional convexity

J.M. Holtzman: Convexity and the Maximum Principle for
Discrete Systems. IEEE Trans. Autom. Control, Vol. AC-11,
No. 1, 1966, pp. 30-35.

J.M. Holtzman: On the Maximum Principle for Nonlinear
Discrete-Time Systems. IEEE Trans Autom. Control, Vol.
AC-11, No. 2, 1966, pp. 273-274.

J.M. Holtzman, H. Halkin: Directional Convexity and the
Maximum Principle for Discrete Systems. SIAM J. Control,
Vol. 4, No. 2, 1966, pp. 263-275.

M.D. Canon, C.D. Cullum Jr., E. Polak: Theory of Optimal
Control and Mathematical Programming. McGraw-Hill, 1970.

Numerical algorithms under assumptions of the existence of a
linear support are discussed in

S. Katz: Best Operating Points for Staged Systems. Ind.
& Eng. Chem. Fundamentals, Vol. 1, No. 4, 1962, pp.
226-240.

M.M. Denn, R. Aris: Green's Functions and Optimal Systems. I, II, III. Ind. & Eng. Chem. Fundamentals, Vol.
4, No. 1, pp. 7-16, No. 2, pp. 213-222, No. 3, pp. 248-257,
1965.

M.M. Denn: Convergence of a Method of Successive Approx-
imations in the Theory of Optimal Processes. Ind. & Eng.
Chem. Fundamentals, Vol. 4, p. 231, 1965.

L.T. Fan, C.S. Wang: The Discrete Maximum Principle.
Wiley, 1964.

CHAPTER 4

COMPUTER ALGORITHM

4.1   Introduction

In this chapter we propose an algorithm for the generalized maximum principle with quadratic support. The algorithm works for
the problems with fixed initial and end points without any constraints on control and states. Moreover the problem must satisfy assumptions of Theorem 222 (nonlinear maximum principle) from
chapter 3.

The chapter is organized as follows. First a general idea of
the algorithm is given. After this the exact algorithm is described. Then results of computations for three examples are
presented. At the end of the chapter some remarks are included.
They concern the experience gained from running the examples and
possibilities of using this algorithm for more complicated
problems. These are mainly the free-end-point problems and
problems with constraints on control and/or state variables.

The computer program for the algorithm was coded in basic FOR-
TRAN IV. The examples were run on an IBM 370/168.

4.2   General idea of the algorithm

To understand better the idea of the algorithm let us discuss
at the beginning intuitively the relation between the multistage and the one-stage problems. Suppose that in the one-
stage problem neither the function $g(u) = x$ nor the upper
boundary $ub(x)$ are known explicitly. We can, however, find
the value of $g(u)$ for any given $u \in U$. We know the required
value of $\bar{x}$ and for any given parameters A, b of the quadratic
support $\pi(x) = x^T A x + b x$ we can find a point x^0 at which
this support supports the upper boundary $ub(x)$. This can be
done by finding a point u^0 which maximizes the Hamiltonian
$H(u,\pi)$ and then calculating the value $g(u^0) = x^0$. Our aim
is to adjust the parameters A and b in such a way that the
support $\pi(x)$ supports the upper boundary $ub(x)$ at the point

$\bar{x}$. Once it is done a value u* maximizing the Hamiltonian is a solution to our problem.

In the multistage fix-end-point problem we know the required value of $\bar{x}_N$ in the last stage. We cannot, however, readily apply the one-stage approach simply by definition of $x = |x_1, x_2, \ldots, x_N|$ and $u = |u_0, u_1, \ldots, u_{N-1}|$. This is so because we do not know optimal values of $x_1^*, x_2^*, \ldots, x_{N-1}^*$ and this way we do not know a value of $\bar{x}$ which is necessary to use the one-stage approach. We can, however, proceed the other way. For any support in the last stage $\pi_N(x_N) = x_N^T A_N x_N + b_N x_N$ we can compute a point x_N at which this support supports the upper boundary $ub_N(x_N)$ as a solution to the generalized maximum principle formulae provided this solution is an optimal solution. This way we can use the concept of one-stage approach for the last stage. For a given support $\pi_N(x_N)$ the value x_N^0 at which this support supports upper boundary $ub_N(x_N)$ is computed as a solution to the generalized maximum principle.

Now, the algorithm will consist of two loops which we shall call the higher loop and the lower loop.

In the higher loop the parameters of the quadratic support in the last stage of the form $\pi_N(x_N) = x_N^T A_N x_N + b_N^T x_N$ are adjusted step by step in order to get the support at the point $\bar{x}_N$. The algorithm attempts to find the appropriate support mainly by changing the values of b_N shifting this way the support along the x_n axis. The exact way of doing it will be explained in the sequel. However, if the rate of convergence of computed values of x_N to $\bar{x}_N$ is too small then the values a_N^{jj} on the main diagonal of the matrix A_N are increased. In the algorithm the rate of convergence is chosen to be 2 i.e. the differences between x_N and $\bar{x}_N$ in each iteration should be at most half of that in the previous iteration. If not, the values a_N^{jj} are multiplied by 10 .

In the lower loop the solution to generalized maximum principle is searched for. This loop is executed for every values of parameters a_N and b_N fixed in the higher level. The lower loop consist at the beginning of computations backwards for $i = N-1, N-2, \ldots, 2, 1, 0$ of the following values

1° a value u_i which maximizes the Hamiltonian $H_i(x_i, u_i, \pi_{i+1})$

2° the value of the vector b_i of the support
$\pi_i(x_i) = x_i^T A_i x_i + b_i x_i$ from the condition

$$\frac{\partial H_i(x_i, u_i, \pi_{i+1})}{\partial x_i} = -\frac{\partial \pi_i(x_i)}{\partial x_i}$$

(the value b_0 is not computed)

and then of computations forwards for $i = 1, 2, \ldots, N$ of

3° the value of state from the equation

$$x_{i+1} = x_i + f_i(x_i, u_i)$$

The values of a_i, $i = 1, 2, \ldots, N$ and b_N during execution of the lower loop once kept constant unless the rate of decreasing of the state variations $|x_i - x_{i-1}|$ is not sufficient. If this is the fact then the values a_i^{jj} on diagonals of matrices A_i, $i = 1, 2, \ldots, N-1$ are increased. The rate of decreasing the state variations is again chosen to be 2 and the values a_i^{jj} are multiplied by 10.

The lower loop is executed unless the state variations are small enough, i.e. $|x_i - x_{i+1}| \leq \epsilon_1$. Then the higher loop computations start again. The higher loop computations stop when the computed value of x_N is close enough to the value of $\bar{x}_N$, say $|x_N - \bar{x}_N| \leq \epsilon$. The value ϵ has to be given, the value ϵ_1 is calculated in the algorithm in the following way

$$\epsilon_1 = [1 + 10 \exp(-\ell)]$$

where ℓ is the higher loop step number.

To start the computations of the algorithm the initial values of
states x_i, $i = 1,2,...,N-1$, are needed. They are provided to
the algorithm from the formula

$$x_i = x_0 + \frac{i}{N}(\bar{x}_N - x_0)$$

The initial values of a_i^{jj}, $i = 1,2,...,N$ are set to 1 and
the initial value of b_N is computed from the formula

$$b_N = -2A_N x_N$$

Let us now proceed to presentation and motivation of formulae for
adjusting parameters b_N. To make the expressions simpler we
drop the subscript N indicating the last stage.

Let us suppose that given a support $\pi(x) = x^T Ax + b^T x$ after
maximizing the Hamiltonian we have found that it supports the upper boundary at a point x^0. The derivative of this support at
x^0 is

$$\frac{\partial \pi(x^0)}{\partial x^T} = 2Ax_0 + b$$

Now we look for a new value of b, say $b*$, that for an optimal
support $\pi*(x) = x^T Ax + b*^T x$ satisfies

$$\frac{\partial \pi*(\bar{x})}{\partial x} = \frac{\partial \pi(x^0)}{\partial x}$$

For differentiable upper boundaries the above condition is
equivalent to the requirement that the derivatives of upper boundaries at the points x^0 and $\bar{x}$ are equal. We can then interpret this condition as an effect of a linear approximation of the
small upper boundary.

Now, substituting for derivatives of $\pi(x)$ and $\pi*(x)$ we have

$$2A\bar{x} + b* = 2Ax^0 + b$$

that is

$$b* = b + 2A\left(x^0 - \bar{x}\right)$$

This leads to a recurrent relation for iterating b

$$b^{(\ell+1)} = b^{(\ell)} + 2A\left(x^{(\ell)} - \bar{x}\right)$$

The above relation was first proposed by Hestenes (1969) and independently by Powell (1969) both under assumption of twice continuously differentiable functions.

To force a better rate of convergence the algorithm increases diagonal values of A. Each change of A needs an appropriate correction of b. Otherwise a jump in the value of x at which the support supports upper boundary will occur. Assuming that $b^{(\ell+1)}$ is in a vicinity of optimal b* the derivatives of the supports with the old values of A and the new values $\tilde{A}$ at the point $x^{(\ell)}$ should be equal, i.e.

$$2\tilde{A}x^{(\ell)} + \tilde{b}^{(\ell)} = 2Ax^{(\ell)} + b^{(\ell)}$$

This gives the corrected value of $\tilde{b}^{(\ell)}$ as

$$\tilde{b}^{(\ell)} = b^{(\ell)} + 2\left(A - \tilde{A}\right)x^{(\ell)}$$

This is the condition proposed earlier by Fletcher (1975).

4.3   The algorithm

Before starting the presentation of the exact algorithm let us define the functions which will be used in it.   These are:

$$\pi_i(x_i) = x_i^T A_i x_i + b_i^T x_i \qquad A_i = \text{diag}\left[a_i^{11}, a_i^{22}, \ldots, a_i^{nn}\right]$$

$$H_i(x_i, u_i, \pi_{i+1}) = r_i(x_i, u_i) - \pi_{i+1}\left[x_i + f_i(x_i, u_i)\right]$$

Now, the algorithm is

1° Read $N, n, m, x_0, \bar{x}_N, \varepsilon$,

2° Set $\qquad a_N^{rjj} := 1 \qquad\qquad\qquad j = 1, 2, \ldots, n$

$\qquad\qquad x_i^{r(0)} := x_i^{(0)} := x_0 + \frac{i}{N}(\bar{x}_N - x_0) \qquad i = 1, 2, \ldots, N$

$\qquad\qquad b_N^r := b_N := -2a_N \bar{x}_N$

$\qquad\qquad \ell = 0$

$\qquad\qquad k = 0$

HIGHER LOOP

3° Set $K_N := \infty$

4° Set $K_i^r := K_i : = \infty \qquad\qquad i = 1, 2, \ldots, N-1$

$\qquad\quad \ell := \ell+1$

$\qquad\quad a_i^{jj} := 1 \qquad\qquad\qquad j = 1, 2, \ldots, n \qquad i = 1, 2, \ldots, N$

LOWER   LOOP

5° Set $k := k+1$

$\qquad\qquad \varepsilon_1 := \varepsilon\left(1 + 10e^{-\ell}\right)$

6° For $i = N-1, N-2, \ldots, 1, 0$ compute

     (i) $u_i^{(k)}$ from $\max\limits_{u_i} H_i\left(x_i^{(k-1)}, u_i, \pi_{i+1}\right)$

     (ii) b_i from $\dfrac{\partial H_i}{\partial x_i} = -\dfrac{\partial \pi_i}{\partial x_i}$ $i \neq 0$

7° For $i = 1, 2, \ldots, N$ compute

     (i) $x_i^{(k)} := x_{i-1}^{(k)} + f_{i-1}\left(x_{i-1}^{(k)}, u_{i-1}^{(k)}\right)$ $x_0^{(k)} := x_0$

     (ii)   if $i \neq N$ $c_i := \left| x_i^{(k)} - x_i^{(k-1)} \right|$

     (iii)   if $i = N$ $c_N := \left| x_N^{(k)} - x_N^{(k-1)} \right|$

8° For $i = 1, 2, \ldots, N$ and $j = 1, 2, \ldots, n$ if $c_i^j \geq K_i/4$

     then $J_{ji} := 1$ else $J_{ji} := 0$

9° For $i = 1, 2, \ldots, N-1$ $K_i := \max\limits_{j} c_i^j$

10° Set $x_i^r := x_i$ $i = 1, 2, \ldots, N$

11° For $i = 1, 2, \ldots, N-1$

     if $K_i \leq K_i^r/2$ go to 15°

12° For $i = 1, 2, \ldots, N-1$ and $j = 1, 2, \ldots, n$

     if $J_{ji} = 1$ then $a_i^{jj} := 10 a_i^{jj}$

13° For $i = 1, 2, \ldots, N-1$ set $K_i^r := K_i$

14° Go to 5°

15° For $i = 1,2,\ldots,N-1$ set $K_i^r := K_i$

16° If $\max_{i} K_i > \varepsilon_1$ go to 5°

HIGHER LOOP

17° If $c_N < K_N/2$ go to 21°

18° Set $b_N := b_N^r$

19° For $j = 1,2,\ldots,n$

if $J_{jN} = 1$ then

(i) $a_N^{jj} := 10a_N^{jj}$

(ii) $b_N^j := b_N^j + 2(a_N^{rjj} - a_N^{jj})x_N^j$

20° Go to 3°

21° $b_N^r := b_N$ $a_N^{rjj} = a_N^{jj}$ $j = 1,2,\ldots,n$ $K_N := \max_{j} c_N^j$

22° $b_N := b_N + 2A_N(x_N - \bar{x}_N)$

23° If $K_N > \varepsilon$ go to 4°

24° STOP

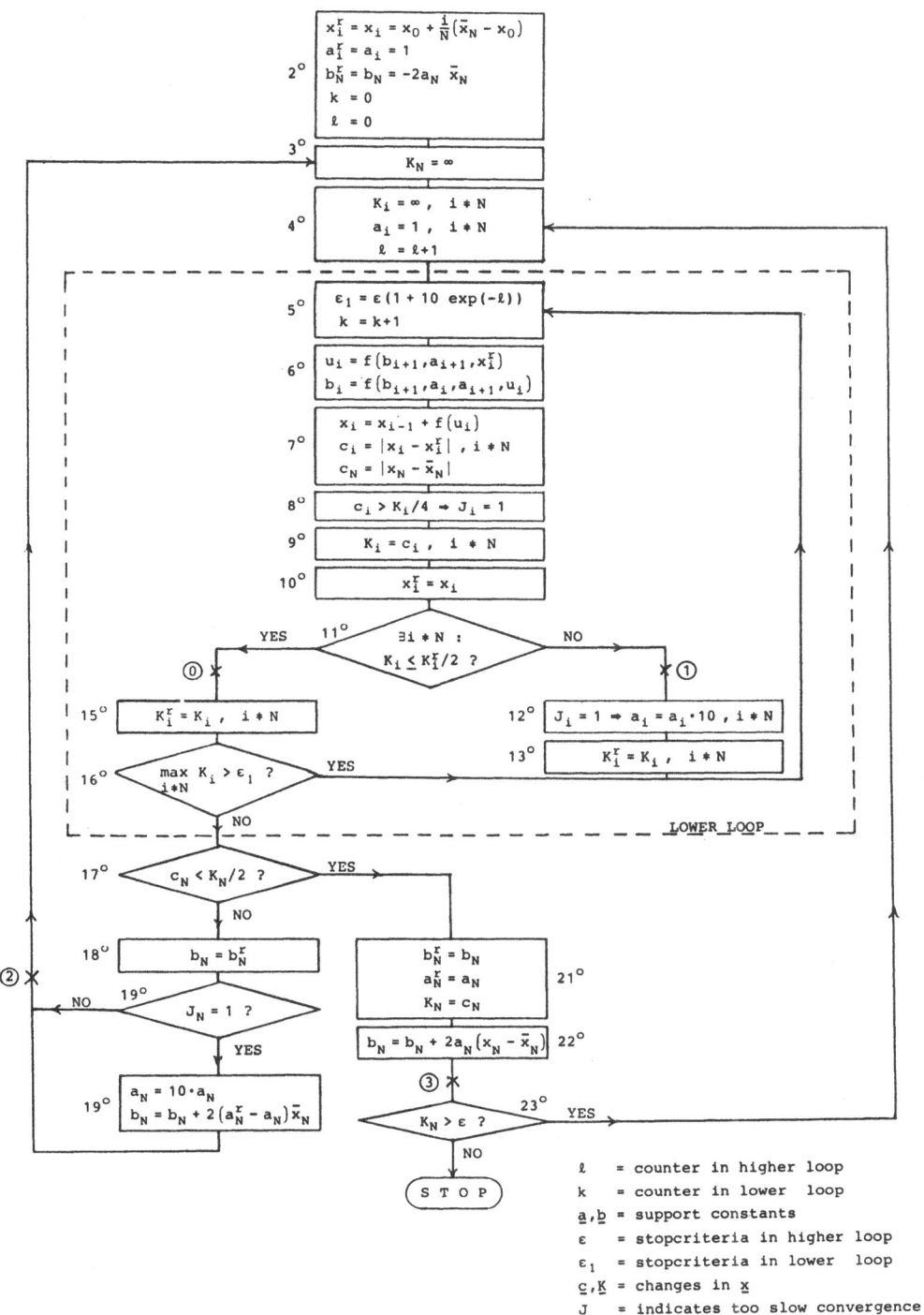

Fig. 4.1   The algorithm.

4.4 Examples

Example 1

This is the example which was used in chapter 3 § 225a, called there Example 1. The problem is

$$\max_{u_0, u_1} \sum_{i=0}^{1} -(u_i)^2$$

subject to

$$x_{i+1} = 2x_i + u_i \qquad i = 0,1$$
$$\bar{x}_0 = 0$$
$$\bar{x}_2 = c$$

The optimal solution was shown to be

$$u_0^* = \tfrac{2}{5} c \qquad u_1^* = \tfrac{1}{5} c \qquad x_1^* = \tfrac{2}{5} c \qquad x_2^* = c$$

with the optimal criterion value

$$R^* = -\tfrac{1}{5} c^2$$

If we want to use the generalized maximum principle with quadratic support then the generalized Hamiltonian is

$$H_i\left(x_i, u_i, a_{i+1}, b_{i+1}\right) = -(u_i)^2 - a_{i+1}\left(2x_i + u_i\right)^2 - b_{i+1}\left(2x_i + u_i\right) \quad i = 1,0$$

The value u_i which maximizes it is equal to

$$u_i^0 = -\frac{b_{i+1} + 4a_{i+1}\, x_i}{2\left(1 + a_{i+1}\right)}$$

Now, from the equation

$$\frac{\partial H_1}{\partial x_1} = -\frac{\partial \pi_1}{\partial x_1}$$

We obtain

$$b_1 = 2b_2 + 4a_2 u_1 + 2(4a_2 - a_1)x_1$$

The results of computation are shown in tables on the next pages. They correspond to 3 values of ε which was used to stop the algorithm. If the absolute value of the difference between x_N obtained from computations and the given value $\bar{x}_N$ is less than ε then the algorithm stops its normal calculation and performs 10 more calculations of the higher loop only with values of a_{ii} from the last normal computation of the loops. In the tables the values of the 10 last calculations are separated by line.

In the computation the values $\bar{x}_0 = 0$ and $\bar{x}_2 = c = 5$ were chosen.

Table 1   CPU time 0.07 s.

Higher loop step number	Number of lower loop steps	Final a_{11} in a step	a_{12}	u_0	u_1	x_1	x_2	R
$x_0 = 0$	Initial values of states					2.500	2.500	$\varepsilon = 0.1$
1	4	10	1	1.591	0.938	1.591	4.119	-3.410
2	2	1	1	1.758	0.795	1.758	4.311	-3.723
3	1	1	10	1.665	0.786	1.665	4.115	-3.389
4	5	100	10	1.765	1.185	1.765	4.716	-4.521
5	5	100	10	1.846	1.165	1.846	4.858	-4.766
6	4	10	100	1.828	1.366	1.828	5.021	-5.206
7	1	10	100	1.875	1.176	1.875	4.927	-4.901
8	1	10	100	1.908	1.118	1.908	4.934	-4.891
9	1	10	100	1.932	1.086	1.932	4.950	-4.911
10	1	10	100	1.950	1.063	1.950	4.963	-4.932
11	1	10	100	1.963	1.047	1.963	4.972	-4.948
12	1	10	100	1.972	1.035	1.972	4.979	-4.961
13	1	10	100	1.980	1.026	1.980	4.985	-4.971
14	1	10	100	1.985	1.019	1.985	4.989	-4.978
15	1	10	100	1.989	1.014	1.989	4.992	-4.984
16	1	10	100	1.992	1.010	1.992	4.994	-4.988
Optimal values				2	1	2	5	-5

Table 2   CPU time 0.19 s.

Higher loop step number	Number of lower loop steps	Final a_{11} in each step	a_{12}	u_0	u_1	x_1	x_2	R
$x_0 = 0$	Initial values of states					2.500	5.000	$\varepsilon = 0.01$
1	4	10	1	1.591	0.938	1.591	4.119	-3.410
2	6	10	1	1.819	0.900	1.819	4.537	-4.117
3	5	100	10	1.809	0.876	1.809	4.494	-4.040
4	5	100	10	1.848	1.041	1.848	4.737	-4.498
5	5	100	100	1.862	0.990	1.862	4.714	-4.447
6	5	100	100	1.891	1.076	1.891	4.859	-4.735
7	5	100	100	1.921	1.088	1.921	4.929	-4.872
8	5	100	1000	1.938	1.048	1.938	4.925	-4.856
9	5	100	1000	1.955	1.053	1.955	4.962	-4.929
10	5	100	10^4	1.965	1.030	1.965	4.960	-4.922
11	5	100	10^4	1.975	1.031	1.975	4.980	-4.961
12	5	100	10^5	1.981	1.018	1.981	4.979	-4.958
13	5	100	10^5	1.986	1.017	1.986	4.989	-4.979
14	3	10	10^6	1.984	1.039	1.984	5.007	-5.016
15	1	10	10^6	1.988	1.017	1.988	4.994	-4.988
16	1	10	10^6	1.991	1.011	1.991	4.994	-4.988
17	1	10	10^6	1.994	1.008	1.994	4.995	-4.991
18	1	10	10^6	1.995	1.006	1.995	4.996	-4.993
19	1	10	10^6	1.997	1.004	1.997	4.997	-4.995
20	1	10	10^6	1.997	1.003	1.997	4.998	-4.996
21	1	10	10^6	1.998	1.002	1.998	4.999	-4.997
22	1	10	10^6	1.999	1.002	1.999	4.999	-4.998
23	1	10	10^6	1.999	1.001	1.999	4.999	-4.998
24	1	10	10^6	1.999	1.001	1.999	4.999	-4.999
Optimal values				2.000	1.000	2.000	5.000	-5.000

Table 3   CPU time 0.32 s.

Higher loop step number	Number of lower loop steps	Final a_{11} in each step	a_{12}	u_0	u_1	x_1	x_2	R
$x_0 = 0$		Initial	values	of	states	2.500	5.000	$\varepsilon = 0.001$
1	6	100	1	1.613	0.888	1.613	4.115	-3.392
2	7	10	1	1.819	0.910	1.819	4.534	-4.112
3	5	100	10	1.805	0.882	1.805	4.492	-4.035
4	7	1000	10	1.848	1.036	1.848	4.732	-4.488
5	7	1000	100	1.862	0.983	1.862	4.707	-4.433
6	7	1000	100	1.894	1.064	1.894	4.853	-4.721
7	7	1000	1000	1.911	1.017	1.911	4.839	-4.685
8	7	1000	1000	1.933	1.053	1.933	4.919	-4.846
9	7	1000	1000	1.953	1.053	1.953	4.960	-4.924
10	5	100	10^4	1.964	1.030	1.964	4.957	-4.917
11	5	100	10^4	1.974	1.031	1.974	4.979	-4.959
12	5	100	10^5	1.980	1.018	1.980	4.977	-4.956
13	5	100	10^5	1.985	1.018	1.985	4.989	-4.978
14	5	100	10^6	1.989	1.010	1.989	4.988	-4.976
15	5	100	10^6	1.992	1.010	1.992	4.994	-4.988
16	5	100	10^7	1.994	1.006	1.994	4.994	-4.987
17	5	100	10^7	1.996	1.006	1.996	4.997	-4.994
18	5	100	10^8	1.997	1.003	1.997	4.997	-4.993
19	5	100	10^8	1.998	1.003	1.998	4.998	-4.997
20	5	100	10^9	1.998	1.002	1.998	4.998	-4.996
21	5	100	10^9	1.999	1.002	1.999	4.999	-4.998
22	3	10	10^{10}	1.999	1.004	1.999	5.001	-5.002
23	1	10	10^{10}	1.999	1.002	1.999	4.999	-4.999
24	1	10	10^{10}	1.999	1.001	1.999	4.999	-4.999
25	1	10	10^{10}	1.999	1.001	1.999	5.000	-4.999
26	1	10	10^{10}	2.000	1.001	2.000	5.000	-4.999
27	1	10	10^{10}	2.000	1.000	2.000	5.000	-5.000
28	1	10	10^{10}	2.000	1.000	2.000	5.000	-5.000
29	1	10	10^{10}	2.000	1.000	2.000	5.000	-5.000
30	1	10	10^{10}	2.000	1.000	2.000	5.000	-5.000
31	1	10	10^{10}	2.000	1.000	2.000	5.000	-5.000
32	1	10	10^{10}	2.000	1.000	2.000	5.000	-5.000
Optimal values				2	1	2	5	-5

Example 2

Consider the following problem

$$\max_{u_0,u_1} \sum_{i=1}^{2} r_i(u_i)$$

$$x_i = x_{i-1} + u_{i-1}$$

$$\bar{x}_0 = 0, \quad \bar{x}_2 = c$$

$$r_0(u_0) = \begin{cases} -10(u_0 + 1)^2 & u_0 \leq -1 \\ 0 & -1 < u_0 < 1 \\ -10(u_0 - 1)^2 & 1 \leq u_0 \end{cases}$$

$$r_1(u_1) = (u_1)^2$$

As we have

$$x_2 = u_0 + u_1 = c$$

the above problem can be solved as a static optimization problem. The Lagrangian is equal to

$$L = r_0(u_0) + r_1(u_1) + \lambda(u_0 + u_1 - c)$$

and we have

$$\frac{\partial L}{\partial u_0} = \begin{cases} -20(u_0 + 1) + \lambda & u_0 \leq -1 \\ \lambda & -1 < u_0 < 1 \\ -20(u_0 - 1) + \lambda & 1 \leq u_0 \end{cases} = 0$$

$$\frac{\partial L}{\partial u_1} = 2u_1 + \lambda = 0$$

$$u_0 + u_1 = c$$

Consider now 3 cases.

1°　$u_0 \leq -1$

Then we have

$$u_0 = \frac{1}{20}\lambda - 1 \qquad\qquad u_1 = -\frac{1}{2}\lambda$$

$$\lambda = -\frac{20}{9}(1+c)$$

and

$$u_0^0 = -\frac{1}{9}(1+c) - 1 \qquad\qquad u_1^0 = \frac{10}{9}(1+c)$$

Now, when $c \geq -1$ the assumption $u_0 \leq -1$ is fulfilled and u_0^0 and u_1^0 give the best controls for this case. The criterion value is equal to

$$R = -\frac{10}{81}(1+c)^2 + \frac{100}{81}(1+c)^2 = \frac{90}{81}(1+c)^2$$

2°　$-1 < u_0 < 1$

The best controls for this case are

$$\left.\begin{array}{l} u_0 = +1 \\[2mm] u_1 = -1+c \end{array}\right\} \quad \text{for} \quad c \leq 0 \qquad\qquad \left.\begin{array}{l} u_0 = -1 \\[2mm] u_1 = 1+c \end{array}\right\} \quad \text{for} \quad c \geq 0$$

and criterion value is equal to

$$R = \begin{cases} (1-c)^2 & \text{for} \quad c \leq 0 \\[3mm] (1+c)^2 & \text{for} \quad c \geq 0 \end{cases}$$

3° $1 \leq u_0$

This case is similar to 1° and we have

$$u_0^0 = \frac{1}{9}(1-c) + 1 \qquad\qquad u_1^0 = -\frac{10}{9}(1-c)$$

$$c \leq 1$$

$$R = \frac{90}{81}(1-c)^2$$

Now we can draw criterion values for all 3 cases

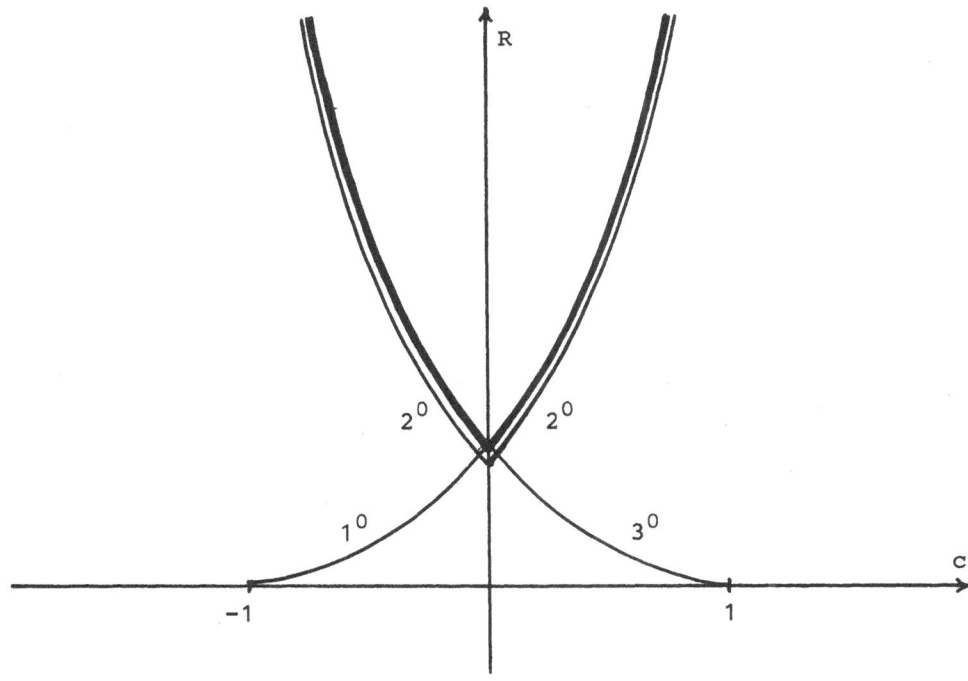

The upper boundary for this problem is marked by thick line. We see that it is not concave. The optimal controls for the problem are

$$u_0^* = \begin{cases} -\frac{1}{9}(1+c) - 1 & c \geq 0 \\ \\ \frac{1}{9}(1-c) + 1 & c \leq 0 \end{cases}$$

$$u_1^* = \begin{cases} \frac{10}{9}(1+c) & c \geq 0 \\ \\ -\frac{10}{9}(1-c) & c \leq 0 \end{cases}$$

$$R^* = \frac{90}{81}(1+|c|)^2$$

The smaller upper boundaries are

$$ub_1(x_1) = \begin{cases} -10(x_1 + 1)^2 & x_1 \leq -1 \\ \\ 0 & -1 < x_1 < 1 \\ \\ -10(x_1 - 1)^2 & 1 \leq x_1 \end{cases}$$

(as $x_1 = u_0$) and

$$ub_2(x_2) = \frac{100}{81}(1+|x_2|)^2$$

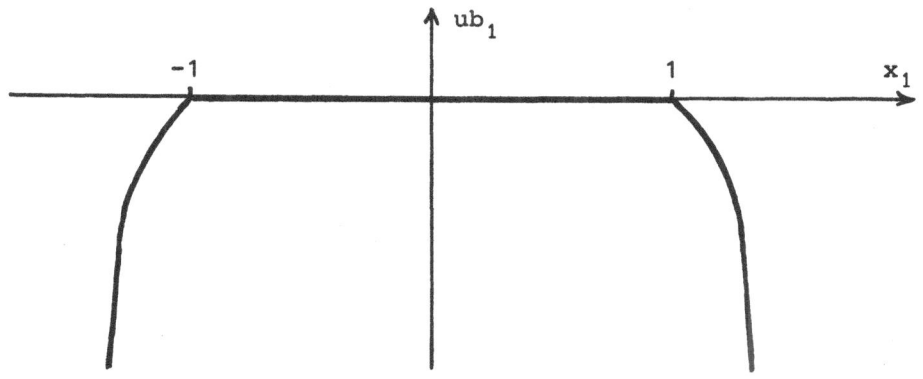

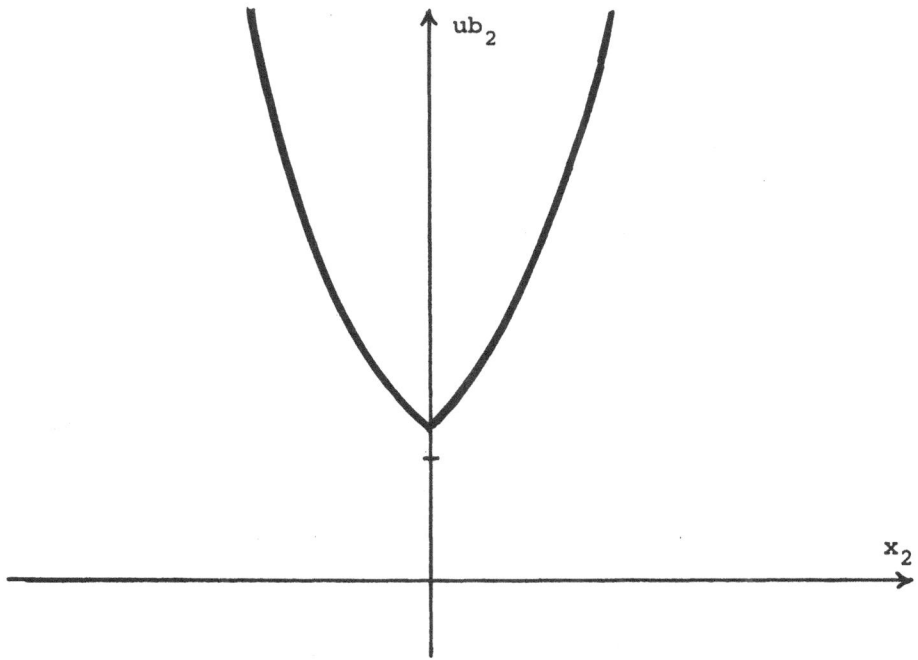

Now, if we want to use the generalized maximum principle with quadratic support, we have

$$H_1(x_1, u_1, a_2, b_2) = (u_1)^2 - a_2(x_1 + u_1)^2 - b_2(x_1 + u_1)$$

As Hamiltonian H_1 is differentiable, then for $a_2 > 1$ it is easy to maximize it by equating the derivative to zero. This gives

$$u_1^0 = -\frac{2a_2 x_1 + b_2}{2(a_2 - 1)}$$

Then we have

$$\frac{\partial H_1}{\partial x_1} = -2a_2(x_1 + u_1) - b_2$$

$$\frac{\partial \pi_1}{\partial x_1} = 2a_1 x_1 + b_1$$

and from the equation

$$\frac{\partial H_1}{\partial x_1} = -\frac{\partial \pi_1}{\partial x_1}$$

we obtain

$$b_1 = b_2 + 2x_1(a_2 - a_1) + 2a_2 u_1$$

At last, the Hamiltonian at stage 0 is

$$H_0(u_0, x_0, a_1, b_1) = r_0(u_0) - a_1(x_0 + u_0)^2 - b_1(x_0 + u_0)$$

and we have

$$\frac{\partial H_0}{\partial u_0} = \begin{cases} -20(u_0 + 1) - 2a_1(x_0 + u_0) - b_1 & u_0 \le -1 \\ \\ -2a_1(x_0 + u_0) - b_1 & -1 < u_0 < 1 \\ \\ -20(u_0 - 1) - 2a_1(x_0 + u_0) - b_1 & 1 \le u_0 \end{cases}$$

Now the value maximizing H_0 is given by

$$u_0 = \begin{cases} -\dfrac{b_1 + 2a_1 x_0 + 20}{2(10 + a_1)} & u_0 \le -1 \\ \\ -\dfrac{b_1 + 2a_1 x_0}{2a_1} & -1 < u_0 < 1 \\ \\ -\dfrac{b_1 + 2a_1 x_0 - 20}{2(10 + a_1)} & 1 \le u_0 \end{cases}$$

From geometrical interpretation we see that one and only one of the above conditions is satisfied.

In the computations the value $\bar{x}_2 = c = 8$ was chosen. The results of the computations are in tables 4, 5, 6.

Table 4   CPU time 0.07 s.

Higher loop step number	Number of lower loop steps	Final a_{11} in each step	a_{12}	u_0	u_1	x_1	x_2	R
$x_0 = 0$		Initial values of states				4.000	8.000	$\varepsilon = 0.1$
1	3	100	10	1.224	7.421	1.224	8.645	54.510
2	3	100	10	−0.378	8.852	−0.378	8.474	78.353
3	3	100	100	−1.591	10.077	−1.591	8.486	98.048
4	3	100	100	−1.909	10.184	−1.909	8.275	95.444
5	1	10	100	−1.965	10.209	−1.965	8.244	94.915
6	1	10	1000	−1.990	10.143	−1.990	8.153	93.085
7	1	10	10^4	−2.003	10.158	−2.003	8.155	93.123
8	1	10	10^4	−2.006	10.093	−2.006	8.087	91.752
9	1	10	10^5	−2.008	10.101	−2.008	8.093	91.874
10	1	10	10^5	−2.007	10.057	−2.007	8.050	91.000
11	1	10	10^5	−2.005	10.031	−2.005	8.026	90.513
12	1	10	10^5	−2.003	10.016	−2.003	8.013	90.252
13	1	10	10^5	−2.002	10.008	−2.002	8.006	90.117
14	1	10	10^5	−2.001	10.004	−2.001	8.003	90.050
15	1	10	10^5	−2.001	10.002	−2.001	8.001	90.019
16	1	10	10^5	−2.000	10.001	−2.000	8.000	90.005
17	1	10	10^5	−2.000	10.000	−2.000	8.000	90.000
18	1	10	10^5	−2.000	10.000	−2.000	8.000	89.998
19	1	10	10^5	−2.000	10.000	−2.000	8.000	89.998
Optimal values				−2	10	−2	8	90

Table 5   CPU time 0.11 s.

Higher loop step number	Number of lower loop steps	Final a_{11} in each step	a_{12}	u_0	u_1	x_1	x_2	R
$x_0 = 0$		Initial values of states				4.000	8.000	$\varepsilon = 0.01$
1	5	1000	10	1.127	7.627	1.127	8.754	58.014
2	5	1000	10	-0.586	9.111	-0.586	8.525	83.003
3	5	1000	100	-1.690	10.262	-1.690	8.572	100.542
4	3	100	100	-1.937	10.217	-1.937	8.280	95.614
5	3	100	100	-1.994	10.140	-1.994	8.146	92.937
6	3	100	1000	-2.011	10.171	-2.011	8.160	93.219
7	1	10	1000	-2.010	10.092	-2.010	8.081	91.634
8	1	10	10^4	-2.010	10.099	-2.010	8.089	91.796
9	1	10	10^4	-2.008	10.055	-2.008	8.047	90.939
10	1	10	10^5	-2.007	10.059	-2.007	8.052	91.044
11	1	10	10^5	-2.005	10.032	-2.005	8.027	90.539
12	1	10	10^6	-2.004	10.034	-2.004	8.030	90.603
13	1	10	10^6	-2.003	10.018	-2.003	8.015	90.309
14	1	10	10^7	-2.003	10.020	-2.003	8.017	90.347
15	1	10	10^7	-2.002	10.011	-2.002	8.009	90.178
16	1	10	10^8	-2.001	10.011	-2.001	8.010	90.200
17	1	10	10^8	-2.001	10.006	-2.001	8.005	90.102
18	1	10	10^8	-2.001	10.003	-2.001	8.002	90.050
19	1	10	10^8	-2.000	10.002	-2.000	8.001	90.023
20	1	10	10^8	-2.000	10.001	-2.000	8.000	90.010
21	1	10	10^8	-2.000	10.000	-2.000	8.000	90.004
22	1	10	10^8	-2.000	10.000	-2.000	8.000	90.001
23	1	10	10^8	-2.000	10.000	-2.000	8.000	90.000
24	1	10	10^8	-2.000	10.000	-2.000	8.000	90.000
25	1	10	10^8	-2.000	10.000	-2.000	8.000	90.000
26	1	10	10^8	-2.000	10.000	-2.000	8.000	90.000
	Optimal values		-2	10		-2	8	90

Table 6   CPU time 0.16 s.

Higher loop step number	Number of lower loop steps	Final a_{11} in each step	a_{12}	u_0	u_1	x_1	x_2	R
$x_0 = 0$		Initial values of states				4.000	8.000	$\varepsilon = 0.001$
1	7	10^4	10	1.117	7.647	1.117	8.764	58.338
2	7	10^4	10	-0.607	9.137	-0.607	8.531	83.494
3	7	10^4	100	-1.699	10.279	-1.699	8.580	100.779
4	5	1000	100	-1.948	10.237	-1.948	8.289	95.815
5	5	1000	100	-1.999	10.143	-1.999	8.144	92.907
6	3	100	100	-2.005	10.077	-2.005	8.071	91.434
7	1	10	100	-2.005	10.041	-2.005	8.036	90.726
8	1	10	1000	-2.005	10.045	-2.005	8.040	90.800
9	3	100	1000	-2.003	10.023	-2.003	8.020	90.397
10	1	10	1000	-2.002	10.013	-2.002	8.011	90.213
11	1	10	10^4	-2.002	10.014	-2.001	8.012	90.238
12	1	10	10^4	-2.001	10.007	-2.001	8.006	90.122
13	1	10	10^5	-2.001	10.008	-2.001	8.007	90.137
14	1	10	10^5	-2.001	10.004	-2.001	8.004	90.070
15	1	10	10^6	-2.001	10.005	-2.001	8.004	90.079
16	1	10	10^6	-2.000	10.002	-2.000	8.002	90.040
17	1	10	10^7	-2.000	10.003	-2.000	8.002	90.045
18	1	10	10^7	-2.000	10.001	-2.000	8.001	90.023
19	1	10	10^8	-2.000	10.001	-2.000	8.001	90.026
20	1	10	10^8	-2.000	10.001	-2.000	8.001	90.013
21	1	10	10^9	-2.000	10.001	-2.000	8.001	90.015
22	1	10	10^9	-2.000	10.000	-2.000	8.000	90.008
23	1	10	10^9	-2.000	10.000	-2.000	8.000	90.004
24	1	10	10^9	-2.000	10.000	-2.000	8.000	90.002
25	1	10	10^9	-2.000	10.000	-2.000	8.000	90.001
26	1	10	10^9	-2.000	10.000	-2.000	8.000	90.000
27	1	10	10^9	-2.000	10.000	-2.000	8.000	90.000
28	1	10	10^9	-2.000	10.000	-2.000	8.000	90.000
29	1	10	10^9	-2.000	10.000	-2.000	8.000	90.000
30	1	10	10^9	-2.000	10.000	-2.000	8.000	90.000
31	1	10	10^9	-2.000	10.000	-2.000	8.000	90.000
Optimal values				-2	10	-2	8	90

Example 3

Let us now consider the multistage compression of a gas problem, see Happel (1958) pp. 65-67. This problem was also solved using dynamic programming by Aris et al. (1960). We have also discussed this problem as the example 1° in chapter 1. To minimize the energy needed to compress a gas isentropically in 3 stages the following function should be maximized

$$R = - \left(\frac{x_1}{x_0}\right)^{\alpha} - \left(\frac{x_2}{x_1}\right)^{\alpha} - \left(\frac{x_3}{x_2}\right)^{\alpha}$$

where $x_0 = p_0$ is an initial pressure, x_i, $i = 1,2$ are intermediate pressures and $x_3 = p_3$ is the final pressure.

The optimal intermediate pressures x_1 and x_2 after first and second stages of compressions may be found using the differential calculus. We have

$$\frac{\partial R}{\partial x_1} = -\alpha \left[\frac{x_1^{\alpha-1}}{x_0^{\alpha}} - \frac{x_2^{\alpha}}{x_1^{\alpha+1}}\right] = 0$$

$$\frac{\partial R}{\partial x_2} = -\alpha \left[\frac{x_2^{\alpha-1}}{x_1^{\alpha}} - \frac{x_3^{\alpha}}{x_2^{\alpha+1}}\right] = 0$$

which after rearranging leads to equations

$$x_1^{2\alpha} = x_0^{\alpha} \, x_2^{\alpha}$$

$$x_2^{2\alpha} = x_1^{\alpha} \, x_3^{\alpha}$$

The solution of this equation is

$$x_1^* = \left(x_0^2 \, x_3\right)^{1/3} = \left(p_0^2 \, p_3\right)^{1/3}$$

$$x_2^* = \left(x_0 \, x_3^2\right)^{1/3} = \left(p_0 \, p_3^2\right)^{1/3}$$

The second partial derivative are

$$\frac{\partial^2 R}{\partial x_1^2} = -\alpha\left[(\alpha-1)\frac{x_1^{\alpha-2}}{x_0^{\alpha}} + (\alpha+1)\frac{x_2^{\alpha}}{x_1^{\alpha+2}}\right]$$

$$\frac{\partial^2 R}{\partial x_2^2} = -\alpha\left[(\alpha-1)\frac{x_2^{\alpha-2}}{x_1^{\alpha}} + (\alpha+1)\frac{x_3^{\alpha}}{x_2^{\alpha+2}}\right]$$

$$\frac{\partial^2 R}{\partial x_1 \partial x_2} = \alpha^2 \frac{x_2^{\alpha-1}}{x_1^{\alpha+1}}$$

and at the point (x_1^*, x_2^*)

$$\frac{\partial^2 R^*}{\partial x_1^2} = -2\alpha^2 \, x_0^{-1/3\alpha-4/3} \, x_3^{1/3\alpha-2/3} < 0$$

$$\frac{\partial^2 R^*}{\partial x_2^2} = -2\alpha^2 \, x_0^{-1/3\alpha-2/3} \, x_3^{1/3\alpha-4/3} < 0$$

$$\frac{\partial^2 R^*}{\partial x_1 \partial x_2} = \alpha^2 \, x_0^{-1/3\alpha-1} \, x_3^{1/3\alpha-1} > 0$$

Now as

$$\frac{\partial^2 R^*}{\partial x^2}\frac{\partial^2 R^*}{\partial x^2} - \left(\frac{\partial^2 R^*}{\partial x_1 \partial x_2}\right)^2 = 3\alpha^4 \, x_0^{-2/3\alpha-2} \, x_3^{2/3\alpha-2} > 0$$

then the matrix of second derivatives is negative definite at (x_1^*, x_2^*) and the function takes maximum at this point.

The problem may be formulated as a control problem

$$\max_{u_0, u_1, u_2} \sum_{i=0}^{2} -(u_i)^{\alpha}$$

subject to

$$x_{i+1} = x_i \, u_i \qquad\qquad i = 0,1$$

$$x_0 = p_0$$

$$x_3 = p_3$$

From earlier calculations the optimal solution to this problem is seen to be

$$u_0^* = \left(\frac{p_3}{p_0}\right)^{1/3}$$

$$x_1^* = \left(p_0^2 \, p_3\right)^{1/3}$$

$$u_1^* = \left(\frac{p_3}{p_0}\right)^{1/3}$$

$$x_2^* = \left(p_0 \, p_3^2\right)^{1/3}$$

$$u_2^* = \left(\frac{p_3}{p_0}\right)^{1/3}$$

with the optimal criterion value

$$R^* = -3\left(\frac{p_3}{p_0}\right)^{\alpha/3}$$

The usual values for α , see Happel (1958) p. 260, are $0 < \alpha < 1$. Then the function $R^*(p_3)$, the greater upper boundary, has the form

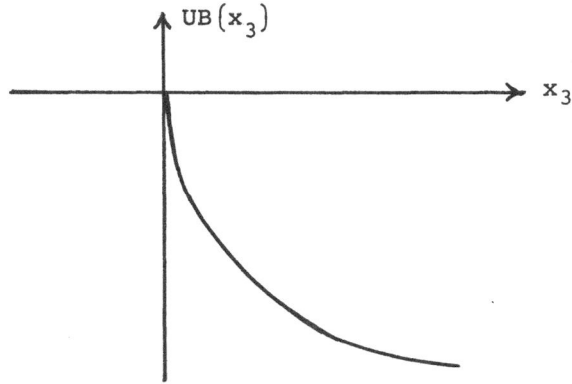

and is not concave. This means that the classical (linear) maximum principle cannot be applied.

In the same way as above we can find

$$UB_2(x_2) = -2\left(\frac{x_2}{x_0}\right)^{\alpha/2}$$

which is also not concave. Also smaller upper boundaries

$$ub_1(x_0,x_1) = -\left(\frac{x_1}{x_0}\right)^{\alpha}$$

$$ub_2(x_1,x_2) = -\left(\frac{x_2}{x_1}\right)^{\alpha}$$

$$ub_3(x_2,x_3) = -\left(\frac{x_3}{x_2}\right)^{\alpha}$$

are not concave.

To use the generalized maximum principle with quadratic Hamiltonian we have

$$H_i(x_i,u_i,a_{i+1},b_{i+1}) = -(u_i)^{\alpha} - a_{i+1}(x_i)^2(u_i)^2 - b_{i+1}\, x_i\, u_i$$

$$i = 2,1,0$$

As the maximum of this Hamiltonian cannot be found analytically the numerical maximization was performed. To do this the golden section method was chosen.

From the equation

$$\frac{\partial H_i}{\partial x_i} = -\frac{\partial \pi_i}{\partial x_i}$$

we get the recursion for b_i

$$b_i = b_{i+1} \, u_i + 2x_i \left[a_{i+1} \left(u_i \right)^2 - a_i \right] \qquad i = 2,1$$

In the tables 7 and 8 the results of computation for $\bar{x}_0 = 1$, $\bar{x}_3 = 8$, $\alpha = 0.29$ and 2 values $\varepsilon = 0.1$ and $\varepsilon = 0.01$ are given. The slow convergence can be observed. This is probably caused by the very flat maximum of the function R. Let us notice that the matrix of the second partial derivative for above values of $\bar{x}_0 = 1$, $\bar{x}_3 = 8$ and $\alpha = 0.29$ is

$$
\begin{bmatrix}
\dfrac{\partial^2 R^*}{\partial x_1^2} & \dfrac{\partial^2 R^*}{\partial x_1 \, \partial x_2} \\[3mm]
\dfrac{\partial^2 R^*}{\partial x_2 \, \partial x_1} & \dfrac{\partial^2 R^*}{\partial x_2^2}
\end{bmatrix}
=
\begin{bmatrix}
0.0514 & 0.0128 \\[3mm]
0.0128 & 0.0128
\end{bmatrix}
$$

which means that in the vicinity of the optimal solution $\left(x_1^*, x_2^* \right)$ any change of the value x_1 produces change of only 5% of $|R - R^*|/|x_1 - x_1^*|$ and any change of x_2 only 1.3% of $|R - R^*|/|x_1 - x_2^*|$.

Referring back to the formulae for the second derivatives it can be seen that their values can be made bigger taking smaller values of x_3. This was verified on a computer and the results are shown in table 9 where much better values of control and states are performed. As the number of lower loop steps in each iteration was equal to 1 and so were the value of a_{ii}, $i = 1,2,3$, then they are neither shown in the table 9 nor in the tables 10 and 11 which show the results for higher number of stages $N = 4$ and $N = 10$ respectively.

Table 7   CPU time 0.14 s.

Higher loop step number	Number of lower loop steps	Final a_{11}	Final a_{12}	a_{13}	u_0	u_1	u_2	x_1	x_2	x_3	R
$x_0 = 1$				Initial values of states				3.333	5.667	8.000	$\varepsilon = 0.1$
1	1	1	1	1	3.317	1.700	1.409	3.317	5.637	7.941	-3.687
2	1	1	1	1	3.303	1.699	1.421	3.303	5.612	7.975	-3.688
3	1	1	1	1	3.290	1.698	1.430	3.290	5.585	7.987	-3.688
4	1	1	1	1	3.275	1.698	1.437	3.275	5.561	7.994	-3.688
5	1	1	1	1	3.260	1.698	1.444	3.260	5.534	7.994	-3.687
6	1	1	1	1	3.244	1.698	1.452	3.244	5.508	7.998	-3.687
7	1	1	1	1	3.234	1.696	1.459	3.234	5.485	8.004	-3.687
8	1	1	1	1	3.219	1.697	1.465	3.219	5.461	7.998	-3.686
9	1	1	1	1	3.202	1.696	1.472	3.202	5.431	7.993	-3.686
10	1	1	1	1	3.190	1.695	1.480	3.190	5.406	8.004	-3.686
11	1	1	1	1	3.174	1.694	1.487	3.174	5.377	7.995	-3.685
Optimal values					2	2	2	2	4	8	-3.668

Table 8   CPU time 0.15 s.

Higher loop step number	Number of lower loop steps	Final a_{11}	Final a_{12}	a_{13}	u_0	u_1	u_2	x_1	x_2	x_3	R
$x_0 = 1$					Initial values of states			3.333	5.667	8.000	$\varepsilon = 0.01$
1	1	1	1	1	3.316	1.700	1.409	3.316	5.637	7.940	-3.687
2	3	10	10	1	3.285	1.698	1.435	3.285	5.577	8.005	-3.688
3	1	10	10	1	3.281	1.698	1.436	3.281	5.571	8.001	-3.688
4	1	10	10	1	3.278	1.699	1.437	3.278	5.568	8.003	-3.688
5	1	10	10	1	3.270	1.700	1.438	3.270	5.558	7.993	-3.687
6	1	10	10	1	3.261	1.701	1.441	3.261	5.548	7.996	-3.687
7	1	10	10	1	3.259	1.702	1.444	3.259	5.545	8.006	-3.688
8	1	10	10	1	3.251	1.703	1.445	3.251	5.535	7.998	-3.687
9	1	10	10	1	3.247	1.703	1.447	3.247	5.530	8.003	-3.687
10	1	10	10	1	3.239	1.704	1.449	3.239	5.520	7.997	-3.687
11	1	10	10	1	3.233	1.705	1.451	3.233	5.513	7.999	-3.687
12	1	10	10	1	3.228	1.706	1.453	3.228	5.507	8.003	-3.687
Optimal values					2	2	2	2	4	8	-3.668

Table 9   CPU time 0.11 s.

Higher loop step number	u_0	u_1	u_2	x_1	x_2	x_3	R
$x_0 = 1$	Initial values of states			1.243	1.487	1.730	$\epsilon = 0.1$
1	1.242	1.194	1.103	1.242	1.482	1.635	-3.146
2	1.240	1.193	1.139	1.240	1.479	1.684	-3.155
3	1.238	1.192	1.158	1.238	1.476	1.709	-3.160
4	1.236	1.191	1.168	1.236	1.472	1.719	-3.161
5	1.234	1.191	1.174	1.234	1.470	1.726	-3.163
6	1.233	1.191	1.178	1.233	1.468	1.728	-3.163
7	1.230	1.190	1.180	1.230	1.464	1.728	-3.163
8	1.229	1.189	1.184	1.229	1.462	1.730	-3.163
9	1.228	1.188	1.186	1.228	1.460	1.730	-3.163
10	1.226	1.188	1.187	1.226	1.457	1.729	-3.163
11	1.224	1.188	1.189	1.224	1.455	1.731	-3.163
Optimal values	1.200	1.200	1.200	1.200	1.441	1.730	-3.163

Table 10   CPU time 0.14 s.

Higher loop step number	u_0	u_1	u_2	u_3	x_1	x_2	x_3	x_4	R
$x_0 = 1$	Initial values of states				1.270	1.540	1.810	2.080	$\varepsilon = 0.1$
1	1.268	1.212	1.174	1.109	1.268	1.537	1.803	1.999	-4.206
2	1.266	1.210	1.174	1.135	1.266	1.532	1.798	2.040	-4.213
3	1.264	1.209	1.174	1.150	1.264	1.528	1.794	2.062	-4.216
4	1.262	1.208	1.174	1.157	1.262	1.525	1.790	2.072	-4.217
5	1.261	1.207	1.174	1.163	1.261	1.521	1.785	2.076	-4.218
6	1.259	1.205	1.174	1.167	1.259	1.517	1.781	2.078	-4.218
7	1.256	1.204	1.174	1.170	1.256	1.513	1.776	2.078	-4.218
8	1.254	1.203	1.174	1.174	1.254	1.509	1.772	2.080	-4.218
9	1.253	1.202	1.174	1.177	1.253	1.506	1.768	2.080	-4.218
10	1.250	1.202	1.174	1.179	1.250	1.503	1.764	2.080	-4.218
11	1.248	1.201	1.174	1.182	1.248	1.499	1.760	2.080	-4.218
Optimal values	1.201	1.201	1.201	1.201	1.201	1.442	1.732	2.080	-4.218

Table 11   CPU time 0.33 s.

Higher loop step number	u_0	u_1	u_2	u_3	u_4	u_5	u_6	u_7	u_8	u_9
1	1.517	1.341	1.254	1.203	1.168	1.145	1.126	1.112	1.100	1.088
2	1.513	1.340	1.254	1.203	1.167	1.145	1.125	1.112	1.100	1.097
3	1.508	1.339	1.254	1.202	1.167	1.144	1.125	1.112	1.100	1.105
4	1.504	1.339	1.254	1.202	1.167	1.144	1.125	1.112	1.100	1.111
5	1.500	1.337	1.254	1.202	1.167	1.144	1.124	1.112	1.101	1.117
6	1.497	1.335	1.254	1.201	1.167	1.143	1.123	1.112	1.102	1.121
7	1.493	1.333	1.254	1.201	1.166	1.142	1.124	1.112	1.101	1.128
8	1.490	1.331	1.254	1.201	1.166	1.141	1.124	1.112	1.101	1.133
9	1.487	1.329	1.254	1.201	1.166	1.141	1.124	1.112	1.101	1.137
10	1.483	1.328	1.254	1.200	1.166	1.141	1.124	1.112	1.102	1.142
11	1.480	1.326	1.254	1.200	1.166	1.140	1.124	1.112	1.102	1.147
Optimal values	1.200	1.200	1.200	1.200	1.200	1.200	1.200	1.200	1.200	1.200

Higher loop step number	x_1	x_2	x_3	x_4	x_5	x_6	x_7	x_8	x_9	x_{10}	R
$x_0 = 0$ Initial values	1.520	2.040	2.560	3.080	3.600	4.120	4.640	5.160	5.680	6.200	$\varepsilon = 0.1$
1	1.517	2.034	2.551	3.069	3.584	4.104	4.620	5.139	5.655	6.151	-10.545
2	1.513	2.027	2.542	3.058	3.568	4.086	4.598	5.115	5.627	6.173	-10.546
3	1.508	2.020	2.533	3.046	3.554	4.068	4.576	5.090	5.600	6.186	-10.547
4	1.504	2.013	2.524	3.034	3.540	4.048	4.554	5.066	5.574	6.193	-10.547
5	1.500	2.006	2.515	3.023	3.526	4.033	4.533	5.042	5.551	6.199	-10.547
6	1.497	1.998	2.505	3.010	3.511	4.013	4.508	5.015	5.525	6.195	-10.547
7	1.493	1.991	2.496	2.999	3.497	3.994	4.489	4.994	5.498	6.201	-10.547
8	1.490	1.984	2.488	2.989	3.485	3.978	4.472	4.975	5.478	6.204	-10.547
9	1.487	1.977	2.479	2.977	3.471	3.961	4.452	4.952	5.452	6.198	-10.547
10	1.483	1.969	2.470	2.964	3.456	3.942	4.431	4.927	5.427	6.198	-10.547
11	1.480	1.962	2.461	2.954	3.444	3.926	4.412	4.906	5.406	6.202	-10.547
Optimal values	1.200	1.440	1.729	2.075	2.490	2.988	3.587	4.304	5.166	6.200	-10.543

4.5 Complementary remarks

The algorithm given in p. 4.3 is only a first proposition for
solving the multistage optimization problems using the generalized maximum principle. The three simple examples presented in
p. 4.4 showed reasonable results of applying it. There is, however, still too little experience to draw any conclusions. Some
new solutions in the algorithm formulation may prove useful in
future applications. Therefore computing more numerical examples
and real cases would be welcomed.

Few constants control the way of computation in the algorithm.
These are:

- the constant 4 in step 8° which decides when the increase of
  values a_i^{jj} may happen,
- the constant 2 in step 11° which decides whether the increase
  of values a_i^{jj} in lower loop should be performed or not,
- the constant 2 in step 17° which decides whether the increase
  of values a_i^{jj} in higher loop should be performed or not.

No bigger investigations of the influence of those paremeters
have been made. From our experience it seems only that forcing
too quick convergence i.e. high values of parameters make the
results worse. Examination of the influence of those parameters
as well as some others as for example multiplication by 10 the
values of a_i^{jj} in steps 12° and 19° or the way of setting the
value ε_1 in step 5° could be the subject of further researches.

As it is seen from the examples the algorithm stops sometimes
too early. 3-5 more iterations give much better results. An
improvement of the stopping condition might be the other subject
of research. One idea is to examine the variations in criterion
values.

The algorithm may be applied to broader class of problems. One important area are the problems with bounds on controls and/or states. These kind of problems may be solved using the above algorithm together with the appropriate penalty functions. The other class of problems are free-end-point problems. To handle them a simple modification could be tried. It consists of assuming that $\pi_N(x_N) = 0$ i.e. $A_N = 0$ and $b_N = 0$. Then only the lower loop should be executed.

4.6 Literature

The idea of iterating support parameter b_N comes from the papers dealing with static problems

M.R. Hestenes: Multiplier and Gradient Methods. JOTA, Vol. 4, No. 5, 303-320, 1969.

M.J.D. Powell: A Method for Nonlinear Constraints in Minimization Problems. In: R. Fletcher (ed.): Optimization. Academic Press, 1969, pp. 283-298.

The computational experience of applying the Hestenes/Powell method is discussed in

A. Miele, P.E. Moseley, A.V. Levy, G.M. Coggins: On the Method of Multipliers for Mathematical Programming Problems. JOTA, Vol. 10, No. 1, 1-33, 1972.

R. Fletcher: An Ideal Penalty Function for Constrained Optimization. J. Inst. Maths. Applics., Vol. 15, 319-342, 1975.

The convergence of Hestenes/Powell and associated methods is discussed e.g. in

R.D. Rupp: On the Combination of the Multiplier Method of Hestenes and Powell with Newton's Method. JOTA, Vol. 15, No. 2, 167-187, 1975.

R.T. Rockafellar: The Multiplier Method of Hestenes and Powell Applied to Convex Programming. JOTA, Vol. 12, 555-562, 1973.

R.T. Rockafellar: Solving a Nonlinear·Programming Problem by Way of a Dual Problem. Symposia Matematica, Vol. XXVII, 1976.

R.T. Rockafellar: Penalty Method and Augmented Lagrangians in Nonlinear Programming. Proc. 5th IFIP Conf. on Optimization Techniques. Rome 1973. Springer-Verlag, 1974.

D.P. Bertsekas: Combined Primal-Dual and Penalty Methods for Constrained Minimization. SIAM J. Control, Vol. 13, No. 3, 521-544, 1975.

D.P. Bertsekas: On Penalty and Multiplier Methods for Constrained Minimization. SIAM J. Control and Optim., Vol. 14, No. 2, 216-235, 1976.

D.P. Bertsekas: Multiplier Methods: A Survey. Automatica, Vol. 12, 133-145, 1976.

D.P. Bertsekas: On the Convergence Properties of Second-Order Multiplier Methods. JOTA, Vol. 25, No. 3, 443-449, 1978.

D.P. Bertsekas: Penalty and Multiplier Methods. In: L.C.W. Dixon, E. Spedicato, G.P. Szegö (eds.): Nonlinear Optimization, Theory and Algorithms. Birkhäuser, Boston 1980, pp. 253-278.

D.P. Bertsekas: Constrained Minimization and Lagrange Multiplier Methods. Academic Press (in preparation).

Algorithms for lower level loop (for linear supports) were discussed in

S. Katz: Best Operating Points for Staged Systems. Ind. &
Eng. Chem. Fundamentals, Vol. 1, No. 4, 226-240, 1962.

L.T. Fan, Ch. S. Wang: The Discrete Maximum Principle.
Wiley, 1964.

M.M. Denn, R. Aris: Green's Functions and Optimal Systems.
I, II, III. Ind. & Eng. Chem. Fundamentals, Vol. 4, No. 1,
7-16, No. 2, 213-222, No. 3, 248-257, 1965.

Convergence of these algorithms was discussed in the above paper
by Denn & Ans and also in

M.M. Denn: Convergence of a Method of Successive Approxi-
mations in the Theory of Optimal Processes. Ind. & Eng.
Chem. Fundamentals, Vol. 4, No. 2, 231-232, 1965.

The solution of the multistage compression of a gas problem was
discussed in

J. Happel: Chemical Process Economics. Wiley (1958).

R. Aris, R. Bellman, R. Kalaba: Some Optimization Problems
in Chemical Engineering. Chem. Engg. Progr. Symp., Series
No. 31, Vol. 56, p. 95 (1960).

CHAPTER 5

CONCLUSIONS AND FURTHER RESEARCH

In this book we have presented a new approach to optimization of the multistage optimization problems called the upper boundary approach.

The classical methods of solving this problem like the maximum principle method and the dynamic programming method are shown to fit smoothly into the new approach. Moreover, using this approach a number of new results have been developed, among these a new, generalized version of the discrete maximum principle is arrived at. The new version does not require the assumption of directional convexity, an assumption which has forbidden the application of the classical discrete maximum principle to some practical multistage optimization problems. An example was given of a simple technical problem of the multistage compression of a gas; it was shown that it did not have the directional convexity property. This and two other simple problems were solved numerically using an algorithm for finding a solution to the generalized maximum principle conditions.

This book contains the thorough discussion of the present status of the works on the upper boundary approach. However, it does not pretend to complete the research in this area. On the contrary, it is felt that many further works should be done to make this approach a flexible tool for solving a variety of practical problems including not only technical or operational systems but also others like economical, biological, ecological, etc. systems. It seems to be important that to proceed in that direction some real life problems should be solved with the theory given in this book. This will enable to see the limitations of the theory and the needs for extending it. This will also contribute to the verification of the algorithm given in chapter 4 and its possible improvement. It will moreover show how some practical difficulties can be overcome when using this approach.

Apart from these more practically-oriented investigations some further theoretical works can proceed. One of them is an extension of the generalized maximum principle for boundary conditions. Many multistage optimization problems have optimal solutions on the boundaries and the only way to deal with them in the present theory is to use the penalty functions techniques. The straight application of the generalized maximum principle for active boundaries is not possible only because of the assumptions for part (iii) of Theorem 222. It seems that this assumption might be released when using some kind of complementary slackness conditions.

During discussions on the results obtained an idea has appeared that the generalized maximum principle could be generalized even more by defining a new function of the form

$$C_i(x_i, u_i, \pi_{i+1}) = r_i(x_i, u_i) - \pi_{i+1}\Big[f_i(x_i, u_i) + x_i\Big]$$
$$- \pi^0_{i+1}(x_{i+1}) + UB_i(x_i)$$

Defining now a saddle-point to C_i as a point $(x^0_i, u^0_i, \pi^0_{i+1})$ satisfying

$$C_i(x^0_i, u_i, \pi^0_{i+1}) \leq C_i(x^0_i, u^0_i, \pi^0_{i+1}) \leq C_i(x^0_i, u^0_i, \pi_{i+1})$$

we could get all the results of Chapter 3 §§ 217-219 and 222-223 from two basic theorems:

1. $(x^*_i, u^*_i, \pi^*_{i+1})$ is a saddle-point to $C_i(x_i, u_i, \pi_{i+1})$.

2. Under assumptions of differentiability

$$\text{(i)} \qquad \frac{\partial C_i(x^*_i, u^*_i, \pi^*_{i+1})}{\partial x_i} = 0$$

$$\text{(ii)} \qquad \frac{\partial C_i(x^*_i, u^*_i, \pi^*_{i+1})}{\partial u_i} = 0$$

$$(iii) \qquad \frac{\partial C_i\left(x_i^*, u_i^*, \pi_{i+1}^*\right)}{\partial \pi_{i+1}} = 0$$

To do it a suitable definition of differentiability with respect to the function π_{i+1} is needed. This requires thorough look at the fundamentals of the theory. Then an application of functional analysis methods would be probably necessary.

A good deal of work may be done on the theoretical investigations of properties of the numerical algorithm such like e.g. convergence and its rate. The argumentation of convergence of the multipliers method given in the literature cited at the end of Chapter 4 could be a good starting point. Combining these results with the conditions for convergence of the lower loop migth give solution to the algorithm convergence problem.

The upper boundary approach has very intuitionally appealing geometrical interpretation. This property could be used for the elaboration of the popular presentation of this theory which could be further developed for lecturing purposes.

The upper boundary approach can be extended to the continuous-time systems. The way of doing it has already been presented by Ravn (1980). This proposition needs mainly a mathematical treatment.

In short, the book documents in specific areas the applicability of the approach, and above a number of possible extensions of great interest are stretched. Thus we think that we have here shown that the Upper Boundary Approach is a fruitful tool for tackling discrete time systems.

Lecture Notes in Control and Information Sciences

Edited by A. V. Balakrishnan and M. Thoma

Lecture Notes in Control and Information Sciences

Edited by A. V. Balakrishnan and M. Thoma